Archimedes'
Revenge

Archimedes' Revenge

The Joys and Perils of Mathematics

□

PAUL HOFFMAN

W · W · Norton & Company · New York · London

Portions of this book appeared in another
form in the *New York Times Magazine* and in
Science Digest.

Published simultaneously in Canada by Penguin Books
Canada Ltd., 2801 John Street, Markham, Ontario
L3R 1B4.
Printed in the United States of America.

The text of this book is composed in Avanta, with
display type set in Avant Garde Book. Composition
and manufacturing by the Haddon Craftsmen.
Book design by Bernard Klein.

First Edition

ISBN: 978-0-393-32775-5

W. W. Norton & Company, Inc., 500 Fifth Avenue,
New York, N. Y. 10110
W. W. Norton & Company Ltd., 37 Great Russell
Street, London WC1B 3NU

1 2 3 4 5 6 7 8 9 0

For Martin Gardner

CONTENTS

Archimedes'
Revenge

INTRODUCTION

There was more imagination in the head of
Archimedes than in that of Homer.
—VOLTAIRE

When Isaac Newton made his famous understatement "If I have seen further than [others], it is by standing upon the shoulders of giants," he surely had in mind Archimedes of Syracuse, the greatest mathematician of antiquity. Archimedes, however, was also a mechanical genius, inventing, among other gadgets, the water snail, or Archimedean screw, a helical pump for raising water for irrigation. Although little is known about Archimedes' life or about his assessment of his own work, most commentators suspect that he valued his theoretical mathematical discoveries more than his practical inventions. Plutarch, for one, writes, "And yet Archimedes possessed such a lofty spirit, so profound a soul, and such a wealth of scientific theory, that, although his inventions had won for him a name and fame for superhuman sagacity, he would not consent to leave behind him any treatise on this subject, but regarding the work of an engineer and every art that ministers to the needs of life as ignoble and vulgar, he devoted his earnest efforts only to those studies the subtlety and charm of which are not affected by the claims of necessity." Other commentators add that even when he dealt with levers, pullies, or other machines, he was seeking general principles of mechanics, not practical applications.

How much Archimedes truly preferred the theoretical to the practical may never be known. It is clear, however, that in his work there is tension between theory and application, a tension that still pervades mathematics twenty-two centuries later.

My aim in this book is to sketch the range and scope of mathematics. I do not pretend that this book is comprehensive. Indeed, it is quirky in its choice of subjects. But it couldn't be otherwise. Mathematics is a discipline practiced in every university in the world, and it is at least as broad a field as biology, in which one researcher tries to understand the AIDS virus while another studies the socialization of wombats.

I approach mathematics as I do a Chinese menu, trying dishes here and there, recognizing both common ingredients and distinctive flavors. After only one Chinese meal, you'll hardly be an expert on Chinese cuisine, but you'll know much more about Chinese food than someone who has never eaten it at all. So it is with mathematics. By dipping into a handful of mathematical topics, you'll not learn everything that's important in mathematics, but you'll have a much better feel for the subject than someone who hasn't taken the plunge.

Many books have been written about the philosophical underpinnings of mathematics, about the extent to which it is the science of certainty, in that its conclusions are logically unassailable. Many other works have rhapsodized at length about the nature of infinity and the beauty of higher dimensions.

Such philosophical and poetic excursions have their place, but they are far from the concerns of most working mathematicians. In this book I give a glimpse of some of the things that mathematicians, pure and applied, actually do.

I also want to counter a misconception: that any result in mathematics can be achieved just by laboriously doing enough computation—that, in other words, if you want to solve a mathematical problem, it's merely a matter of doing enough arithmetic. Granted, you and I lack the computational skills to attack complex mathematical problems, but we suspect that those in the know—those who understand mathematical symbols—can grind out an answer to almost any problem if they choose to. After all, we are taught to believe that mathematics is syllogistic, that deducing a mathematical result is as straightforward as drawing the conclusion "Socrates is mortal" from the premises "All men are mortal" and "Socrates is a man." If only mathematics were that simple!

One of my aims is to convey a sense of the limits of mathematical knowledge. In each area of mathematics that we examine, I'll point out what is known and not known. Sometimes our knowledge is limited because a field is young and not many mathematicians have devoted themselves to it. In other cases, little is known because the problems are extraordinarily difficult. In still other cases, there are more fundamental reasons for the mathematician's limited knowledge; it can be shown that the problems are simply immune to quick mathematical solutions.

Mathematics is full of surprises. Number and shape are among humanity's oldest concerns, and yet much about them is still not understood. What could be simpler than the concept of a prime number—an integer greater than 1 like 3, 5, 17, or 31 that cannot be evenly divided by any integer other than 1 and itself? The ancient Greeks knew that the supply of primes is inexhaustible, but no one knows whether the supply of twin primes—pairs of primes, such as 3 and 5, that differ by 2—is infinite, too. No one knows whether there's an infinite number of perfect numbers, integers like 6 that are equal to the sum of all of their divisors except, of course, the integer itself (in this case, 3, 2, and 1). And no one knows if a perfect number can be odd. Paul Erdös, the great Hungarian number theorist who is a master of proving basic theorems about primes—at the age of eighteen, he came up with a celebrated proof that there is always a prime between every integer greater than 1 and its double—believes that mathematicians are nowhere near understanding the integers, let alone other kinds of numbers. "It will be another million years, at least," says Erdös, "before we understand the primes."

The mathematical understanding of shape is no more advanced. In two dimensions, many questions remain unanswered about what shapes can be used to tile a surface, given certain basic constraints. The three-dimensional analogue of the tiling problem, the packing of shapes as densely as possible into a given space, is not solved for many basic shapes. Lack of theoretical knowledge, however, need not always stand in the way of pragmatists, as the designer Ronald Resch demonstrated when he built a three-and-a-half-story Easter egg.

With fundamental questions about number and shape still unsettled, it is no wonder that there is much disagreement and confusion about what the computer—a very complex mathematical tool—can and cannot do. I have tried to stay clear of mushy metaphysical issues about the nature of man and machine in favor of presenting what little is known about the theoretical limits on computing. I discuss the surprising power of Turing's universal computing machine—a strip of paper divided into cells. And I look at a probable limitation: computer scientists think they'll be able to prove that a certain class of simple-sounding computational problems—including that of the traveling salesman who wants to choose the shortest route between a bunch of cities—can never be efficiently solved by machine (or mathematician). Moving from theory to practice, I look at the efforts of Hans Berliner and Danny Hillis to design, respectively, a chess-playing machine and a general-purpose computer that takes the idea that "two heads are better than one" to an amazing extreme. It is much too early to see completely how these efforts will pan out, but both machines are already outperforming traditional computers in certain areas.

The traveling-salesman problem is definitely mathematical in nature, yet traditional mathematical attacks have proved to be of little help in its solution. A similar situation, I'll show, arises in the design of a voting system or a method of apportioning representatives. Mathematics can't help in an absolute sense. Indeed, mathematics demonstrates the theoretical futility of creating a perfectly democratic voting system. But short of perfect democracy, mathematics points the way to a fair voting system and a fair method of congressional apportionment.

Legend has it that Archimedes, in a fit of rage, composed an insanely difficult numerical problem about grazing cattle. His revenge was felt for twenty-two hundred years, until 1981, when the problem was finally disposed of by a fledgling supercomputer. The cattle problem is somewhat contrived. Yet the frustration generations of mathematicians felt in the face of Archimedes' revenge resembles that caused by simpler mathematical problems that arise more naturally. The revenge that mathematics itself has wrought shows no sign of abating.

A formidable student at Trinity
Solved the square root of infinity;
It gave him such fidgets
To count up the digits
He chucked math and took up divinity.

—ANONYMOUS

1

NUMBERS

In ancient Greece, there were no Social Security numbers, no phone numbers, no censuses, no postelection polls, no statistical data, and no 1099 forms to fill out. The world was not yet digitized, and yet numbers were foremost in the minds of the Greek intelligentsia. Indeed, in the sixth century B.C. Pythagoras of Samos made a kind of religion out of the study of numbers, regarding them not as mere instruments of enumeration but as sacred, perfect, friendly, lucky, or evil. The branch of mathematics called number theory—the study of the properties of integers—began with the ancient Greeks and is still flourishing today.

The next three chapters are devoted to number theory. In them, I emphasize that some of the oldest and most elementary-sounding problems are still unsolved. *Why* they are unsolved is unclear, but *that* they are is a fact that looms large and that should dispel any notion that mathematics is some kind of rote activity. Number theory used to be regarded as the purest branch of mathematics; it seemed to have no application to the real world. Recently, however, it has become a powerful tool in cryptography. But, as I discuss in the fourth chapter, "The Cryptic Case of a Swarthy Stranger," there are legendary codes that to this day defy mathematical analysis.

1

666 AND FRIENDS

Michael Friedman, now an undergraduate at the Massachusetts Institute of Technology, was a cocky high school senior from Brooklyn when he won third place in the 1985 Westinghouse Science Talent Search. For his award-winning project, he didn't want to dirty his hands with brine shrimp, fruit flies, or flatworms. And he didn't want to tackle just any age-old theoretical question. No, he chose to confront a problem so old that it could well be the oldest unsolved problem in mathematics, a problem that confounded the ancient Greeks and everyone since: Is there an odd perfect number?

Pythagoras and his cronies saw perfection in any whole number that equaled the sum of all its divisors (except the number itself). The first perfect number is 6. It's evenly divisible by 1, 2, and 3, and it's also the sum of 1, 2, and 3. The second perfect number is 28. Its divisors are 1, 2, 4, 7, and 14, and they add up to 28. That much the Greeks knew, but try as they did, they could not find an odd perfect number.

The perfection of the numbers 6 and 28, biblical commentators have observed, is reflected in the structure of the universe: God created the world in six days, and the moon orbits Earth every twenty-eight days. Yet it is the numbers themselves, not any connection to the empirical world, that makes them perfect. Saint Augustine put it this way: "Six is a number perfect in itself, and not because God created all things in six days; rather the inverse is true; God created all things in six days because this number is perfect. And it would remain perfect even if the work of the six days did not exist."

"This whole area of mathematics is high-level doodling," says Peter Hagis, Jr., a mathematics professor at Temple University. "I got involved with perfect numbers out of idle curiosity, because it's probably the oldest unsolved problem. It is, perhaps, a trivial pursuit, yet the problem is so old that it's not considered a complete waste of time to work on it. If the problem had first been brought up five years ago, it would be totally uninteresting."

Perfection in any realm should be hard to come by, and the even perfect numbers are no exception, but at least they are known to exist. Thirty of them have been discovered, the largest being the 130,000-digit monstrosity $2^{216,090} (2^{216,091} - 1)$. Perhaps a thirty-first perfect number will not turn up. For more than twenty-three hundred years mathematicians have known that there are infinitely many primes (numbers that can be evenly divided only by themselves and by 1). But in that same period of time, they have not been able to determine whether perfect numbers are also inexhaustible.

I would have been happy to interview Michael Friedman over Cokes at the Russian Tea Room or the Four Seasons, but he preferred that we meet in his principal's office at Stuyvesant High School, a Manhattan enclave for math and science jocks. Einstein, it is rumored, could barely add or subtract but could do higher mathematics in his sleep. The same might be said of Michael. The simple act of choosing a time for our meeting was a minor ordeal because the whiz kid was not adept at

Perfect Number		Number of Digits
1. $2^1(2^2-1)$ =	6	1
2. $2^2(2^3-1)$ =	28	2
3. $2^4(2^5-1)$ =	496	3
4. $2^6(2^7-1)$ =	8,128	4
5. $2^{12}(2^{13}-1)$ =	33,550,336	8
6. $2^{16}(2^{17}-1)$ =	8,589,869,056	10
7. $2^{18}(2^{19}-1)$ =	137,438,691,328	12
8. $2^{10}(2^{11}-1)$ =		19
9. $2^{60}(2^{61}-1)$ =		37
10. $2^{88}(2^{89}-1)$ =		54
11. $2^{106}(2^{107}-1)$ =		65
12. $2^{126}(2^{127}-1)$ =		77
13. $2^{520}(2^{521}-1)$ =		314
14. $2^{606}(2^{607}-1)$ =		366
15. $2^{1,278}(2^{1,279}-1)$ =		770
16. $2^{2,202}(2^{2,203}-1)$ =		1,327
17. $2^{2,280}(2^{2,281}-1)$ =		1,373
18. $2^{3,216}(2^{3,217}-1)$ =		1,937
19. $2^{4,252}(2^{4,253}-1)$ =		2,561
20. $2^{4,422}(2^{4,423}-1)$ =		2,663
21. $2^{9,688}(2^{9,689}-1)$ =		5,834
22. $2^{9,940}(2^{9,941}-1)$ =		5,985
23. $2^{11,212}(2^{11,213}-1)$ =		6,751
24. $2^{19,936}(2^{19,937}-1)$ =		12,003
25. $2^{21,700}(2^{21,701}-1)$ =		13,066
26. $2^{23,208}(2^{23,209}-1)$ =		13,973
27. $2^{44,496}(2^{44,497}-1)$ =		26,790
28. $2^{86,242}(2^{86,243}-1)$ =		51,924
29. $2^{112,048}(2^{112,049}-1)$ =		79,502
30. $2^{216,090}(2^{216,091}-1)$ =		130,100

The Thirty Perfect Numbers

converting high school time—"third period" and "fifth period"—into the hours and minutes that the rest of us go by. Yet once we did get together, the gawky phenom proved to be an articulate delight.

"I had to do a paper last year for a math teacher," Michael told me. "I knew about the problem of odd perfect numbers. It interested me because it's so simple, but no one has found the answer." First, Michael looked into the history of perfect numbers.

The ancients knew of only four perfect numbers, the first four: 6, 28, 496, and 8,128. Euclid recognized—and only the Greek gods know how—that these four numbers are generated by the formula $2^{n-1}(2^n - 1)$ for n equals 2, 3, 5, and 7. The computations are as follows:

For $n = 2$, $2^1(2^2 - 1) = 2\,(3) = 6.$
For $n = 3$, $2^2(2^3 - 1) = 4\,(7) = 28.$
For $n = 5$, $2^4(2^5 - 1) = 16\,(31) = 496.$
For $n = 7$, $2^6(2^7 - 1) = 64\,(127) = 8{,}128.$

Euclid saw that in all four computations $2^n - 1$ was a prime (3, 7, 31, and 127). This observation inspired him to prove a powerful theorem: the formula $2^{n-1}(2^n - 1)$ generates an even perfect number whenever $2^n - 1$ is a prime.

Euclid's proof got the theory of perfect numbers off to a roaring start, but the nearsightedness of other mathematicians made progress slow. Many fine minds thought they saw patterns in the numbers where none existed. If they had looked a little further, they would have seen that the patterns were illusory.

The ancients observed that each of the first four perfect numbers ended in a 6 or an 8. Moreover, the final digits were seen to alternate 6, 8, 6, 8. It was therefore assumed that the final digit would always be a 6 or an 8 and that they would continue to alternate. The fifth perfect number, which the ancients did not know, does end in a 6. But the sixth perfect number, alas, ends in a 6, too, which breaks the alternating pattern. The ancients were right, however, about the last digit's always being a 6 or an 8. Today, mathematicians can study thirty perfect numbers—over seven times more than the ancients could—but they have yet to find a pattern to the terminal 6s and 8s.

The ancients also observed that the first perfect number has one digit, the second perfect number has two digits, the third has three, and the fourth has four. They assumed, therefore, that the fifth perfect number would have five digits. Seventeen centuries after Euclid, the fifth perfect number was discovered, and it weighed in at a whopping eight digits: 33, 550, 336. The numbers continued to grow rapidly, the next three being 8,589,869,056; 137,438,691,328; and 2,305,843,008,139,952,-128.

Euclid's proof that $2^n - 1(2^n - 1)$ will yield a perfect number whenever $2^n - 1$ is a prime says nothing about which integral values of n will make $2^n - 1$ a prime. Since the first four values of n that make $2^n - 1$ a prime are the first four prime numbers $(2, 3, 5, 7)$, it might be assumed that whenever n is a prime, $2^n - 1$ is also a prime. Well, let's try the fifth prime number, 11. For $n = 11$, $2^n - 1$ is 2,047, which is not a prime (it's the product of 23 and 89). The truth is that n must be a prime for $2^n - 1$ to be a prime, but n being prime does not in itself make $2^n - 1$ a prime. Indeed, for most prime values of n, $2^n - 1$ is *not* a prime.

Numbers generated by the expression $2^n - 1$ are now known as Mersenne numbers, after a seventeenth-century Parisian monk, Marin Mersenne, who took time out from his monastic duties for number theory. On account of Euclid's formula, everytime a new *prime* Mersenne number is discovered, a new perfect number is automatically known. In 1644, Mersenne himself stated that the three Mersenne numbers $2^{13} - 1$, $2^{17} - 1$, and $2^{19} - 1$ are primes (8,191; 131,071; and 524,287). The monk also claimed that the huge Mersenne number $2^{67} - 1$ would prove to be a prime. This bold claim went unchallenged for more than a quarter of a millennium.

In 1903, at a meeting of the American Mathematical Society, Frank Nelson Cole, a Columbia University professor, rose to deliver a paper modestly entitled "On the Factoring of Large Numbers." Eric Temple Bell, the historian of mathematics, recorded what happened: "Cole—who was always a man of very few words—walked to the board and, saying nothing, proceeded to chalk up the arithmetic for raising 2 to the sixty-seventh power. Then he carefully subtracted 1 [getting the 21-digit monstrosity 147,573,952,589,676,412,927]. Without a word he moved over to a clear space on the board and multiplied out, by longhand,

$$193,707,721 \times 761,838,257,287.$$

"The two calculations agreed. Mersenne's conjecture—if such it was—vanished into the limbo of mathematical mythology. For the first and only time on record, an audience of the American Mathematical Society vigorously applauded the author of a paper delivered before it. Cole took his seat without having uttered a word. Nobody asked him a question."

Some two thousand years after Euclid proved that his formula always yields even perfect numbers, the eighteenth-century Swiss mathematician Leonard Euler proved that the formula will yield *all* the even perfect numbers. The problem of odd perfect numbers can then be put another way: Are there any perfect numbers not generated by Euclid's formula?

To see what recent progress had been made, the young Michael Friedman plowed through back issues of *Mathematics of Computation, Journal of Number Theory, Acta Arithmetica*, and a host of other periodicals rarely found on coffee tables. He even consulted Richard Guy's formidable classic, *Unsolved*

Problems in Number Theory, which discusses not only perfect numbers but also dozens of other arcane subjects: "almost superperfect numbers," "friendship graphs," "graceful graphs," "greedy algorithms," "loopy games," "Davenport-Schnitzel sequences," "quasi-amicable numbers," "sociable numbers," and "untouchable numbers."

Michael learned that number theorists, frustrated by the intractability of the problem, have proved all sorts of things about what an odd perfect number must be like if one exists. It must be evenly divisible by at least eight different prime numbers, of which the largest must exceed 300,000 and the second largest must exceed 1,000. If an odd perfect number is not divisible by 3, it must be divisible by at least eleven different prime numbers. Moreover, an odd perfect number must leave a remainder of 1 when divided by 12 or a remainder of 9 when divided by 36.

What are we to make of these results? The more constraints there are on an odd perfect number, the less likely it is that one exists. Indeed, in 1973, using constraints like these and aided by a computer, Peter Hagis definitively proved that there is no odd perfect number below 10^{50}. Since 1973, Michael read in Guy's book, other number theorists have "gradually pushed the bound below which an odd perfect number cannot exist, to above 10^{200}, though there is some skepticism about the latter proofs."

Since no less an authority than Guy questioned these proofs, Michael decided to tackle the lower-bound problem anew. With an IBM PC and a list of constraints, including some from India rarely mentioned in the literature, Michael demonstrated that there are no odd perfect numbers below 10^{79} that have eight prime divisors (which is the minimum number of prime divisors an odd perfect number could have).

"In my paper," said Michael, "I just quoted Guy's statement that previous proofs [of high lower bounds for odd perfect numbers] were suspect. When I got into the Westinghouse finals, I decided to review the other proofs but couldn't find any reason why they were suspect. So I phoned Guy up, and he told me that mathematicians don't like proofs that are done by computers, because you never know: Did the guy make a little mistake coding it? Did the computer have a glitch?"

Even supposing that the computer's number crunching checks out (on, say, another computer), the proofs themselves are often so long and complex that no one other than their author has gone through them step by step. Only Hagis's proof (all eighty-three pages of it!) has been thoroughly dissected by other mathematicians and pronounced valid.

Michael grinned broadly. "My proof," he said proudly, "is also suspect." Either the folks at Westinghouse didn't catch on or didn't care. "As far as I know," Michael noted, "no one really reviewed my paper."

On the basis of his paper and other supporting material, Michael was one of 40 Westinghouse finalists selected from a pool of 1,100 applicants. The 40 were summoned to Washington, where they would be whittled down to 10 winners. "Once you're in Washington," Michael explained, "it's almost not based on your paper. You're interviewed by a bunch of scientists. They'd ask, 'How do you measure the distance between the earth and the sun? How do you measure the height of the Washington Monument?' One girl said, 'With a tape measure.' One of the scientists was wearing a tie with half the periodic table on it. He asked everyone about the periodic table. Some people noticed the tie and read off the answers. I didn't, so I had to remember about the number of protons in oxygen and about electron orbitals."

When Michael added, "We were also questioned by a psychiatrist," I winced. "That's what everyone does when I tell them about the psychiatrist. He asked people about their family life. Westinghouse wants to identify future Nobel Prize winners. That's their big thing. They want future Nobel winners to be in the top ten." Michael explained that five past Westinghouse finalists (there are forty a year, and the competition has been going on for forty-four years) have won the Nobel Prize, but of these five, only one was in the top ten. Michael patiently explained to me that Westinghouse was doing worse than random. (The random selection each year of ten from forty would have resulted in 1.25 Nobel Prize winners in the top ten. Leave it to mathematicians to conceive of a quarter of a scientist going to Stockholm.) To improve their record, the psychiatrist was evidently brought in to spot the germ of Nobel Prize–winning personalities.

"My adviser," Michael went on, "wrote in my application that I wouldn't let go of a problem, that I was very stubborn. So the psychiatrist spent the entire fifteen minutes asking me about being stubborn. 'How stubborn are you? Do you think that being stubborn would ever hurt you in your later life? Do you ever refuse suggestions just because you originally argued against them?' "

Since Michael made it into the top ten, perhaps stubbornness is part of the Nobel laureate profile. Unfortunately for Westinghouse (and for Michael), there is no Nobel Prize in mathematics or computer science. If he wants one, he may have to play with brine shrimp after all.

But abandoning perfect numbers may even be good for Michael's health. Others who have studied them too long have been inexorably drawn into the numerical mysticism of the ancients. Michael Stifel and Peter Bungus were mathematicians in the Renaissance who failed to unravel the mystery of perfect numbers; Stifel asserted incorrectly that all perfect numbers except 6 are divisible by 4, and Bungus made a false claim about their terminal digits. Having toyed with perfection, Stifel and Bungus turned to the opposite quality—evilness—which they found in 666, the notorious number of the beast.

Wallace John Steinhope, the physicist in Paul Nathan's story "Newton's Gift," is obsessed with the idea that Isaac Newton and other scientific luminaries of yesteryear must have squandered inordinate amounts of time on tedious mathematical computations. Imagine poor Newton suffering endless delays in the discovery of gravity because of a simple error in arithmetic! When Steinhope invents a knapsack-size time machine, he decides to go to the England of 1666—Newton's golden year and, coincidentally, the final year of that century's great plague—and bestow on Newton a pocket calculator. Steinhope's motive, of course, is to emancipate "Newton's mighty brain from tedium."

Newton, however, is afraid of the calculator, particularly its glowing red digital display: "As the Lord is my savior, is it a creation of Lucifer? The eyes of it shine with the color of his domain."

"You cannot deny your own eyes," Steinhope responds. "Let me *show* you it works. I'll divide two numbers for you with just the punch of a few buttons." Steinhope entered, at random, 81,918 divided by 123. When the answer lit up, Newton fell to his knees and started to pray. Then he got up, grabbed a hot poker from the fireplace, and swung it at Steinhope, who barely escaped back to the space-time coordinates of today.

Newton's violent reaction can be explained by Steinhope's unfortunate choice of numbers: 81,918 divided by 123 happens to be 666, the number of the beast. The religious Newton was aghast to see the calling card of the fallen archangel pulsate before him in eerie red light. It was this brush with the devil, it is said, that inspired Newton to write theological tracts.

Although this clever story is fictitious, it is true in spirit to Newton's fascination with the occult and the supernatural. Newton penned more than 1,300,000 words on biblical and theological subjects. He wrote extensively about interpreting the language of prophets, and he was undoubtedly familiar with biblical prophecies involving the nefarious number 666. Because other men of science and mathematics have been caught up in the mysticism of 666, it is worth exploring how this number got such a bad rep.

In the Middle Ages, a group of Jewish scholars known as cabalists had an ingenious answer to religious heretics who pointed to apparent inconsistencies, trivialities, and non sequiturs in Scripture. Much of the Old Testament, the cabalists claimed, is in code. That's why Scripture may seemed muddled. But when the code is broken, everything will make sense and divine truth will be revealed. The chief method of decryption was *gematria:* a word or phrase is converted into a number by taking all the letters, substituting a predetermined numerical value for each one, and computing the sum of these numbers. The word or phrase is thought to be related to other words or phrases that yield the same sum.

In Genesis 18:2, for example, Abraham looks up, "and lo! three men stood by him," but the men are not identified. The cabalists used gematria to discover that these men were the archangels Michael, Gabriel, and Raphael. If the letters in the original Hebrew for "and lo! three men" are replaced by their numerical equivalents, they sum to 701, as do the letters in "these are Michael, Gabriel, and Raphael." By similar cryptomathematical methods, the cabalists were able to answer the question, posed in Deuteronomy 30:12, "Who shall go up for us to heaven?" A combination of letters from the beginnings and ends of these words in Hebrew yields the same numerical sum as the Hebrew words for circumcision and Jehovah, implying that God saw circumcision as a passport to heaven. This numerical manipulation of Scripture fostered an interest in mathematics among Jewish scholars.

Christian theologians were quick to adopt the analytical methods of the cabalists. The New Testament itself actually promotes the practice of finding the correspondence between names and numbers, and this is where 666 first entered the picture. Revelation 13:11 warns of an evil force: "And I beheld another beast coming up out of the earth; and he had two horns like a lamb, and he spake as a dragon." Seven verses later we learn that the beast is a man associated with the number 666: "Here is wisdom. Let him that hath understanding count the number of the beast: for it is the number of a man; and his number is six hundred threescore and six." But who is the man? The above verse is an open invitation to apply gematria to people's names in order to identify the beast.

The beast is the Antichrist, or false Messiah. In biblical times, the false Messiah was thought to be the Roman Empire, which challenged God's rule by establishing a kind of pagan religion, complete with emperor worship and its own priesthood. Biblical commentators have suspected that the beast was Nero, the Roman emperor, but it takes much manipulation to squeeze 666 out of his name. If Nero's name is written in Greek (Neron), the title Caesar appended, and the combination Neron Caesar then transliterated into Hebrew and the letters assigned their numerical equivalents, the enumeration of his name is 666.

In any event, the enigmatic description of the beast as a man whose number is 666 has given generations of numerologists much to mull over. In the sixteenth century, even mathematicians got into the act. Michael Stifel was a German monk who dabbled in algebra and number theory. He was one of the first to use the symbols + and − for addition and subtraction. He slipped a peculiar interpretation of the number of the beast into a classic book on algebra. Determined to impugn the character of Pope Leo X, Stifel put His Holiness's name through contortions.

He spelled out the X as DECIMUS (the Latin word for "tenth") and then changed the U to V, in the spirit of the Romans, to get DECIMVS. From LEO DECIMVS, he picked out the letters that are Roman numerals—L, D, C, I, M, and V—and, for good measure, threw in the X from Leo X. Now, substituting numbers for the Roman numerals, Stifel computed the numerical value of the name: L(50) + D(500) + C(100) + I(1) + M(1,000) + V(5) + X(10) = 1,666.

Oops! A thousand too much. So, thought Stifel, the M, whose value is 1,000, must stand for *mysterium* ("mystery"). By removing the mystery from the group of letters, he got 666 exactly. With this discovery, he renounced his monastic vows and became a follower of Martin Luther.

Stifel could have achieved the same result with fewer dubious contortions if he had focused on the Roman numerals in one of the pope's Latin titles, Vicarius Filii Dei: V(5) + I(1) + C(100) + I(1) + U(5) + I(1) + L(50) + I(1) + I(1) + D(500) + I(1) = 666.

Be that as it may, Stifel managed to achieve what he wanted. Angered by his treasonous discovery, papists threatened to kill him. In 1522, he took refuge in Luther's own house. Luther was glad to have a new convert but told him to forget the numerological hogwash. Stifel did not take the advice but proceeded to comb the Bible for clues to when the world would end. He convinced himself that doomsday was October 18, 1553, and he delivered sermons on the coming of the end until he was arrested. As the day came near, his parishioners spent their savings on good eating and drinking. When they woke up on October 19 and the world was still intact, they wanted to kill their deceiver and would have done so had Luther not intervened. Two death threats in one lifetime were enough for Stifel, so he gave up prophesying and devoted himself fully to mathematics. He went on to become one of the outstanding German algebraists of the sixteenth century.

I should add that Stifel's interpretation of the number of the beast did not go unchallenged. His contemporary Peter Bungus, author of the 700-page *Numerorum Mysterium* (The Mystery of Numbers), managed to foist the number on Luther himself. Take the name Martin Luther and Latinize the surname to get MARTIN LUTERA. Now let the letters from *A* to *I* represent the numbers from 1 to 9 (considering *I* and *J* interchangeable, as was the custom then), the letters *K* to *S* represent the numbers from 10 to 90 (by multiples of 10) and the letters from *T* to *Z* represent the numbers from 100 to 700 (by multiples of 100). With this connection between letters and numbers, Bung saw that M(30) + A(1) + R(80) + T(100) + I(9) + N(40) + L(20) + U(200) + T(100) + E(5) + R(80) + A(1) = 666. Imagine that!

The Bible provides much inspiration for recreational mathematics beyond 666. When a number is used in the Bible that is not a nice round number like 100 or 1,000, it is there because the ancients considered that number to have mystical significance. In general, a particular number acquired occult status if it was found to have certain elegant but simple arithmetic properties, often involving the sum or the product of a string of consecutive integers. For example, in the Gospel according to John (21:11), Jesus and his disciples have a successful fishing trip on the Sea of Tiberias. When they haul in the catch, they find 153 fish: "Simon Peter went up, and drew the net to land full of great fishes, an hundred and fifty and three: and for all there were so many, yet was not the net broken." What is special mathematically about 153? Ponder this before I give the show away.

To begin with, $153 = 1 + 2 + 3 + 4 + 5 + 6 + 7 + 8 + 9 + 10 + 11 + 12 + 13 + 14 + 15 + 16 + 17$. In other words, it is equal to the sum of all the integers from 1 to 17.

But there is more to the magic of 153. It can be expressed in another fundamental way: $153 = 1 + (1 \times 2) + (1 \times 2 \times 3) + (1 \times 2 \times 3 \times 4) + (1 \times 2 \times 3 \times 4 \times 5)$. Contemporary mathematicians would write this equation more economically: $153 = 1! + 2! + 3! + 4! + 5!$ When a number is followed by an exclamation point, you are supposed to take the product of all the integers from 1 to the number itself. This operation is called taking the factorial of the number.

Somewhere along the line, a savant discovered that if you sum the cubes of the digits in 153, you get back 153. To put it simply, $153 = 1^3 + 5^3 + 3^3$. Then, in 1961, according to the mathematics writer Martin Gardner, Phil Kohn of Yoqne'am, Israel, informed the iconoclastic British weekly *New Scientist* that 153 lies dormant in every third number. I leave it to you to figure out what Kohn told *New Scientist,* but here's a hint: Take any multiple of three. Sum the cubes of its digits. Take the result, and sum the cube of its digits. Keep doing this indefinitely.

Let's turn to another number in the Bible, 220. In Genesis

32:14, Jacob gives Esau 220 goats ("two hundred she goats and twenty he goats") as a gesture of friendship. But why 220? The Pythagoreans had identified particular numbers as "friendly," and 220 was the first of these. The concept of a friendly number was based on their idea that a human friend is a kind of alter ego. Pythagoras once said, "A friend is the other I, such as are 220 and 284." What is so special mathematically about these two numbers?

It turns out that 220 and 284 are each equal to the sum of the proper divisors of the other. (Proper divisors are all the numbers that divide evenly into a number, including 1 but excluding the number itself.) The proper divisors of 220 are 1, 2, 4, 5, 10, 11, 20, 22, 44, 55, and 110. Sure enough, $284 = 1 + 2 + 4 + 5 + 10 + 11 + 20 + 22 + 44 + 55 + 110$. The proper divisors of 284 are 1, 2, 4, 71, and 142, and they sum to 220.

In spite of the ancient interest in friendly numbers, a second pair (17,296 and 18,416) was not discovered until 1636, by Pierre de Fermat. By the middle of the nineteenth century, many able mathematicians had searched long and hard for pairs of friendly numbers, and sixty had been found. But it was not until 1866 that the second-*smallest* pair, 1,184 and 1,210, was discovered, by a sixteen-year-old boy.

Modern mathematicians have extended the concept of friendliness from pairs to triplets. In a friendly triplet, the sum of the proper divisors of any of the numbers equals the sum of the other two numbers. This is so with 103,340,640; 123,228,-768; and 124,015,008. Another friendly triplet is 1,945,330,-728,960; 2,324,196,638,720; and 2,615,631,953,920. But such numbers do not look friendly to me. Indeed, notes Joseph Madachy, the great recreational mathematician, friendly triplets "are not easy to find. The numbers in this last set have 959, 959 and 479 divisors, respectively."

Mathematicians, heeding the old adage that safety comes in numbers, are not ones to refrain from taking a good thing too far. Someone decided to see what happens when you take a number, sum the proper divisors, then sum the proper divisors of that sum, and so on, ad nauseam. Well, what happens? Most of the time, nothing interesting, but once in a blue moon you get back the original number somewhere along the way. Take 12,496. Its proper divisors are 1, 2, 4, 8, 11, 16, 22, 44, 71, 88, 142, 176, 284, 568, 781, 1,136, 1,562, 3,124, and 6,248. Add 'em up, and you get 14,288. Sum the proper divisors of 14,288, and you get 15,472. (If you don't believe me, try it yourself!) Repeat this procedure two more times, and you get 14,536 and 14,264, respectively. Now, the proper divisors of 14,264 are 1, 2, 4, 8, 1,783, 3,566, and 7,132. Sum the seven divisors, and you get, lo and behold, 12,496. If you have time to kill, try the same thing starting with the number 14,316. You'll get the number back—after twenty-eight rounds!

2

ARCHIMEDES' REVENGE

When Srinivasa Ramanujan, the great Indian mathematician, was ill with tuberculosis in a London hospital, his colleague G. H. Hardy went to visit him. Hardy, who was never good at initiating conversation, said to Ramanujan, "I came here in taxi-cab number 1729. That number seems dull to me, which I hope isn't a bad omen."

"Nonsense," replied Ramanujan. "The number isn't dull at all. It's quite interesting. It's the smallest number that can be expressed as the sum of two cubes in two different ways." (Somehow, Ramanujan had immediately recognized that 1729 $= 1^3 + 12^3$ as well as $9^3 + 10^3$.)

Ramanujan, who died in 1920 at the age of thirty-two, was a number theorist, a peculiar breed of mathematician who studies the properties of whole numbers. Number theory is one of the oldest areas of mathematics and, in a sense, the simplest. Numbers, of course, are the universal building blocks of mathematics, yet many fundamental questions about them are still unanswered.

In the third century B.C., Apollonious of Perga could not have known what was in store for him—and for generations of mathematicians—when he innocently improved on Archimedes' work on large numbers. "I'll show you who knows about large numbers," Archimedes thought, and for revenge he reportedly concocted a computational problem about grazing cattle whose solution requires numbers so large that it was not solved until recently. Moreover, it was solved not by man but by machine: the fastest computer in the world.

The posing of the insanely difficult cattle problem is but one of many incredible exploits that made Archimedes a legend in his own time. When the Roman general Marcellus blockaded the harbor of Syracuse, Sicily, in 212 B.C. the king of the city, Hieron, called on Archimedes, a relative, to expel the sixty enemy ships. Archimedes had recently discovered the lever (the occasion for his celebrated statement "Give me a place to stand, and I will move the earth"), and he combined levers and pulleys to build huge cranes that hefted the invading ships out of the harbor. The cranes had help in battle from catapults and from a system of convex mirrors that focused sunlight onto the ships, setting them on fire. The Roman fleet was devastated. Marcellus said, "Let us stop fighting this geometrical monster who uses our ships like cups to ladle water from the sea."

For three years, Archimedes held off the enemy. Then, one night, when the Syracusans were preoccupied with a religious celebration, Roman soldiers scaled the walls and opened the gates. As Marcellus' troops rushed in, he said to his men, "Let no one dare lay a violent hand on Archimedes. This man shall be our personal guest."

When one of Marcellus' men found Archimedes in a courtyard, drawing geometric figures in the sand, he disobeyed his orders and drew his sword. "Before you kill me, my friend," Archimedes pleaded, "pray let me finish my circle." The soldier did not wait. As Archimedes lay dying, he said, "They've taken away my body, but I shall take away my mind."

In keeping with Archimedes' wish, his tombstone was engraved with a sphere inscribed in a cylinder—the symbol of his proud discovery that the volume of a sphere is two-thirds the volume of the smallest cylinder that encloses it.

How much of this legend is true? Archimedes was undoubtedly a mechanical genius. There is good evidence that he designed powerful catapults that could hurl fifty-pound shot 300 feet. But recent investigations into the history of technology rule out the possibility that he constructed cranes capable of snatching enemy vessels from the sea. The basis of this myth may have been his invention of a cranelike apparatus for lifting his own (stationary) ships onto land.

Many scientific luminaries, including Galileo Galilei, and the French naturalist Georges-Louis Leclerc, comte de Buffon, were taken by the idea that Archimedes used a mirror to burn enemy ships, much as a child would use a magnifying glass to ignite paper. In theory, such a mirror could be constructed, but it would need a variable focal length that would keep the sun's rays focused on a moving target; that rules out ordinary mirrors. (In 1747, Buffon claimed to have used a complex mirror to set wood on fire at a distance of 150 feet and to melt lead 10 feet closer.) In any event, Archimedes would not have taken the trouble to build a special mirror, because a simple and highly effective incendiary weapon was known at the time: pots of naphtha mixed with a chemical that spontaneously ignites on contact with water were hurled at the enemy ship.

The picturesque story of Archimedes' death may well be true, although one must be suspicious of the words attributed to him. The great Roman orator Cicero came across Archimedes' tomb in 75 B.C. and found it engraved with a cylinder circumscribing a sphere.

What of the cattle problem? Was it really first posed by Archimedes? Whether or not Archimedes really dreamed it up in a fit of pique, he is known to have worked on the problem, so it is at least twenty-two hundred years old.

"Compute, O friend," the problem begins, "the host of the oxen of the sun, giving thy mind thereto, if thou hast a share of wisdom. Compute the number that once grazed upon the plains of the Sicilian isle Trinacria [Sicily itself] and that were divided according to color into four herds, one milk-white, one black, one yellow and one dappled. The number of bulls formed the majority in each herd and the relations between them were":

1. White bulls = yellow bulls + $(1/2 + 1/3)$ black bulls.
2. Black bulls = yellow bulls + $(1/4 + 1/5)$ dappled bulls.
3. Dappled bulls = yellow bulls + $(1/6 + 1/7)$ white bulls.
4. White cows = $(1/3 + 1/4)$ black herd.
5. Black cows = $(1/4 + 1/5)$ dappled herd.
6. Dappled cows = $(1/5 + 1/6)$ yellow herd.
7. Yellow cows = $(1/6 + 1/7)$ white herd.

"If thou canst give, O friend," the problem continues, "the number of bulls and cows in each herd, thou art not all-knowing nor unskilled in numbers but not yet to be counted among the wise." Stripped to its mathematical essentials, the problem so far is to solve seven equations that involve eight unknowns (four groups of bulls specified by color and four groups of cows of corresponding color). It turns out that these equations are not hard to solve. Indeed, they admit infinitely many solutions, the smallest involving a total herd of 50,389,082 cattle, a number that could comfortably graze on Sicily's 6,358,400 acres.

Archimedes, however, did not stop there. He made the problem much more difficult by imposing two additional constraints on the number of bulls:

8. White bulls + black bulls = a square number.
9. Dappled bulls + yellow bulls = a triangular number.

"When thou hast then computed the totals of the herd, O friend," the problem concludes, "go forth as conqueror, and rest assured that thou art proved most skilled in the science of numbers."

By introducing the idea of triangular numbers and square numbers, Archimedes' cattle problem was drawing on the work of Pythagoras. In the sixth century B.C., Pythagoras and his followers had represented numbers as patterns of dots arranged as triangles, squares, or other geometric figures. Numbers like 3, 6, and 10 were called triangular numbers because they could be represented by dots that formed triangles:

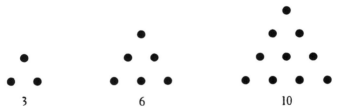

The number of fish Simon pulled from the sea, 153, is also a triangular number. By the same token, numbers like 4, 9, and 16 were called square numbers because they could be represented by dots arranged in squares:

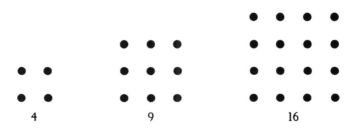

Lest you think that the ancients spent long hours doodling in order to figure out whether a particular number could be represented by a particular geometric dot pattern, you should know there was a purely numerical way of finding this out. All triangular numbers can be generated by summing consecutive integers (starting with 1); thus, $3 = 1 + 2, 6 = 1 + 2 + 3$, and $10 = 1 + 2 + 3 + 4$. All square numbers can be generated by squaring integers: $4 = 2 \times 2, 9 = 3 \times 3$, and $16 = 4 \times 4$.

With the restrictions on the bulls involving triangular and square numbers, the cattle problem became so thorny that no real progress was made for two thousand years. In 1880, a German investigator demonstrated, after tedious computation, that the smallest herd satisfying all eight conditions was a 206,545-digit number that began 776. Archimedes may have been diabolical, but he was certainly not realistic: such a herd could never have fit on the small island of Sicily. As one number theorist put it, "A sphere with a radius equal to the distance from the Earth to the Milky Way could contain only a small part of the animals even if they were the smallest microbes—nay, even were they electrons."

But lack of realism has never been known to impede mathematical research. Two decades later, in 1899, a civil engineer in Hillsboro, Illinois, and a few friends formed the Hillsboro Mathematics Club, devoted to finding the other 206,542 digits. After four years of calculations, they concluded that they had found the 12 rightmost digits and 28 more of the leftmost ones, although subsequent work shows that they goofed in two instances. Another six decades passed before the first complete solution was found, by three Canadians who used a computer, but they never published it. In 1981, all 206,545 digits were finally revealed to the world, when forty-seven pages of printouts from a Cray 1 supercomputer at Lawrence Livermore National Laboratory were reproduced in small type in the *Journal of Recreational Mathematics*.

Back then, the Cray 1 was the fastest computer in the world. Cray supercomputers are expensive machines—the latest version costs $20 million—and laboratories and companies do not buy them in order to solve age-old problems in number theory. They are purchased for use in designing new pharmaceuticals, prospecting for oil, cracking Soviet codes, creating flashy special effects in Hollywood films, and simulating space-based weapons.

Nevertheless, thorny computational problems from the annals of number theory are often given to supercomputers in order to make sure that they're functioning properly. The virtue of such problems is that the solutions, even when previously unknown, can easily be verified by plugging them back into the equations. Archimedes' cattle problem was solved when Lawrence Livermore was testing its new Cray 1. In ten minutes, the supercomputer had found the 206,545-digit solution and double-checked its arithmetic.

It seems only fair to close with a problem Archimedes tackled that *we* might conceivably solve. Hieron had given a goldsmith a known quantity of gold (call its weight W) to make into a crown. When Hieron received the crown, he asked Archimedes to determine whether it contained all the gold or whether the smith had stolen some of it and replaced it with a cheaper metal. According to Vitruvius, the celebrated Roman architect of the first century B.C., "While Archimedes was turning the problem over, he chanced to come to the place of bathing, and there, as he was sitting down in the tub, he noticed that the amount of water which flowed over the tub was equal to the amount by which his body was immersed. This indicated to him a method of solving the problem, and he did not delay, but in his joy leapt out of the tub, and, rushing naked toward his home, he cried out in a loud voice that he had found what he had sought. For as he ran he repeatedly shouted in Greek, *heurika, heurika* [Eureka, eureka, or I have found, I have found]."

What did he find? Archimedes realized that since gold is the densest metal, a pure gold crown of weight W would have a slightly smaller volume than an adulterated gold crown of the same weight. He filled a container to the brim with water and dropped gold of weight W into it. Then he collected the water that overflowed, which was equal in volume to the gold. Next he filled another container with water and dropped the crown under inspection into it. Sure enough, it displaced a larger volume of water, proving that King Hieron had been ripped off by the villainous goldsmith.

3

PRIME PROSTITUTION

Atomism—the belief in *atoma*, "things that cannot be cut or divided"—guided the ancient Greeks in their study not only of matter but also of numbers. Euclid and his contemporaries recognized that certain whole numbers, among them 2, 3, 5, 7, and 11, are essentially indivisible. Called prime numbers, they can be evenly divided only by themselves and the number 1. The numbers that are not prime—4, 6, 8, 9, 10, and so on—have additional divisors. These numbers are said to be composite because each is uniquely "composed" of primes. For example, $4 = 2 \times 2, 6 = 2 \times 3, 8 = 2 \times 2 \times 2, 9 = 3 \times 3$, and $10 = 2 \times 5$.

In September 1985, when the Chevron Geosciences Company in Houston was checking out a new supercomputer, called a Cray X-MP, it identified the largest prime number known to man (or machine) after more than three hours of doing 400 million calculations per second.

Some twenty-three hundred years ago Euclid proved that there are infinitely many primes, but no one has yet found a pattern to them or an efficient formula for generating them. With no pattern to go on, it is no small feat to find a new largest known prime, and news of such a discovery spreads

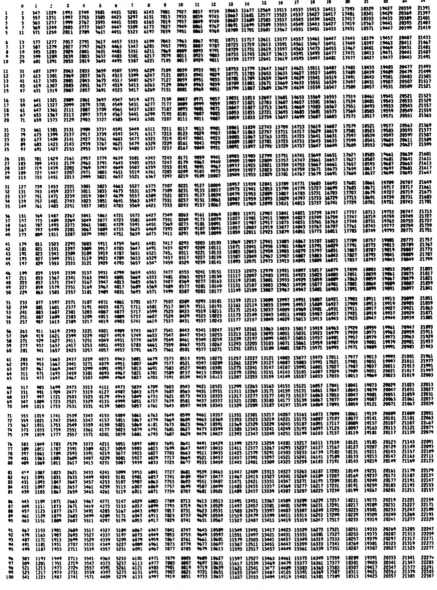

The First 2,500 Primes

From D.N. Lehmer, List of prime numbers from 1 to 10,006,721, Carnegie Institution of Washington. Publication No. 165, Washington, D.C., 1914.

quickly not only through the mathematics community but through the world at large. Walter Cronkite had a soft spot for prime-number stories, and National Public Radio still does.

The record-setting prime that Chevron's computer hit upon weighs in at 65,050 digits. This 65,050-digit whopper is a Mersenne number; it is equal to the number 2 raised to the 216,091st power minus 1. To list all the digits would require 30 pages of this book. "We just happened to crunch enough numbers to come up with a new prime," a Chevron vice-president told the press. "It's my responsibility to get the machine up and running and make sure we have a good one and not a lemon. The results are interesting . . . but they are certainly not going to help me find oil."

Like the search for an odd perfect number, the search for still larger primes, and the probing of their properties, is part of number theory. Number theory is deceptively simple. Its major theorems can often be stated in terms that anyone can understand, but the proofs—when they are known—may require deep, complex mathematics. In 1742, for example, the Prussian-born mathematician Christian Goldbach conjectured that every even number greater than 2 is the sum of two primes. On this analysis, $4 = 2 + 2, 6 = 3 + 3, 8 = 3 + 5, 10 = 5 + 5$, and so on. With the aid of computers, number theorists have decomposed all even numbers up to 100 million into the sum of two primes, but they have not been able to prove that Goldbach's simple conjecture is universally true. And it is not for lack of trying. Over the past two and a half centuries, many of the ablest mathematicians have put their mind to it.

Of all branches of mathematics, number theory has traditionally been the most removed from physical reality. Seemingly abstract results in other esoteric areas of mathematics have been put to good use in physics, chemistry, and economics. This is not true of most results in number theory. If a proof of Goldbach's conjecture were found tomorrow, mathematicians would rejoice but physicists and chemists would not know how to apply the result, if indeed it has any application. Consequently, the contemplation of prime numbers has been regarded as mathematics at its purest, mathematics unadulterated by application. A few centuries ago, this kind of purity earned number theory the appellation "the queen of mathematics."

Today, however, there is trouble in the palace. The purest subjects, the prime numbers, are prostituting themselves in the name of national security. Some of the best codes our government reportedly uses depend on prime numbers. The basis of these codes, in which letters are converted to numbers, is a neat mathematical fact: some computational procedures are comparatively easy to execute but monstrously hard to undo. For example, it is extremely easy for a computer to find the product of two 100-digit primes. But given that 200-digit product, it is immensely difficult to recover those prime divisors (unless, of course, they're told to you). The implications of this for cryptography are mind-boggling. A person who is able to encode messages need not be able to decode them. To encode a message, he need know only the 200-digit product. But to decode the message, he has to know the two prime divisors; knowledge of the product isn't enough.

Such a code is called public key cryptography because it can be used in a highly public way. If I want to receive secret messages, I simply publish the 200-digit number (and an explanation of how it's used for encryption). Then anyone who wishes can send me a coded message. By keeping the two prime divisors to myself, I am the only one who can readily decode the message. The only reason this cryptosystem works, however, is that number theorists have not yet figured out how to factor huge composite numbers into their component primes.

"This kind of cryptosystem," says Carl Pomerance, a noted

prime-number theorist at the University of Georgia, "is an application of ignorance. Because of the codes, more people are involved in number theory. The more mathematicians who knock their heads against the factoring problem [finding the prime divisors] and don't succeed, the better the codes are." The success of this cryptosystem depends on number theory in another way: sophisticated mathematical methods must be used to identify the 100-digit primes that are multiplied together.

Now that prime numbers are at the forefront of cryptography, I want to take stock of what is known (and not known) about them. Euclid long ago proved that the supply of primes is inexhaustible. His 2,300-year-old proof is still a paradigm of mathematical simplicity and elegance.

Assume, said Euclid, that there is a finite number of primes. Then one of them, call it P, will be the largest. Now consider the number Q, larger than P, that is equal to the number 1 plus the product of the consecutive whole numbers from 1 to P. In other words, $Q = 1 + 1 \times 2 \times 3 \ldots \times P$. From the form of the number Q, it is obvious that no integer from 2 to P divides evenly into it; each division would leave a remainder of 1. If Q is not prime, it is evenly divisible by some prime larger than P. On the other hand, if Q is prime, Q itself is a prime larger than P. Either possibility implies the existence of a prime larger than the largest prime. This means, of course, that the concept of "the largest prime" is a fiction. But if there's no such beast, the number of primes must be infinite.

Mathematicians have long dreamed of finding a formula with which, by plugging in integral values of n from 0 to infinity, one could generate all prime numbers. Leonhard Euler, the mathematical phenom of the eighteenth century, played around with the seductively simple formula $n^2 + n + 41$. For $n = 0$, the formula yields the prime number 41; for $n = 1$, the prime number 43; for $n = 2$, the prime number

47. Indeed, as n takes on successive values from 0 to 39, Euler's formula yields nothing but primes. But for $n = 40$, the formula suddenly fails. The result, 1,681, is 41 squared.

n	n^2+n+41	Outcome	n	n^2+n+41	Outcome
0	41	PRIME	32	1097	PRIME
1	43	PRIME	33	1163	PRIME
2	47	PRIME	34	1231	PRIME
3	53	PRIME	35	1301	PRIME
4	61	PRIME	36	1373	PRIME
5	71	PRIME	37	1447	PRIME
6	83	PRIME	38	1523	PRIME
7	97	PRIME	39	1601	PRIME
8	113	PRIME	40	1681	COMPOSITE
9	131	PRIME	41	1763	COMPOSITE
10	151	PRIME	42	1847	PRIME
11	173	PRIME	43	1933	PRIME
12	197	PRIME	44	2021	COMPOSITE
13	223	PRIME	45	2111	PRIME
14	251	PRIME	46	2203	PRIME
15	281	PRIME	47	2297	PRIME
16	313	PRIME	48	2393	PRIME
17	347	PRIME	49	2491	COMPOSITE
18	383	PRIME	50	2591	PRIME
19	421	PRIME	51	2693	PRIME
20	461	PRIME	52	2797	PRIME
21	503	PRIME	53	2903	PRIME
22	547	PRIME	54	3011	PRIME
23	593	PRIME	55	3121	PRIME
24	641	PRIME	56	3233	COMPOSITE
25	691	PRIME	57	3347	PRIME
26	743	PRIME	58	3463	PRIME
27	797	PRIME	59	3581	PRIME
28	853	PRIME	60	3701	PRIME
29	911	PRIME	61	3823	PRIME
30	971	PRIME	62	3947	PRIME
31	1033	PRIME	63	4073	PRIME

n	n^2+n+41	Outcome	n	n^2+n+41	Outcome
64	4201	PRIME	83	7013	PRIME
65	4331	COMPOSITE	84	7181	COMPOSITE
66	4463	PRIME	85	7351	PRIME
67	4597	PRIME	86	7523	PRIME
68	4733	PRIME	87	7697	COMPOSITE
69	4871	PRIME	88	7873	PRIME
70	5011	PRIME	89	8051	COMPOSITE
71	5153	PRIME	90	8231	PRIME
72	5297	PRIME	90	8413	COMPOSITE
73	5443	PRIME	92	8597	PRIME
74	5591	PRIME	93	8783	PRIME
75	5741	PRIME	94	8971	PRIME
76	5893	COMPOSITE	95	9161	PRIME
77	6047	PRIME	96	9353	COMPOSITE
78	6203	PRIME	97	9547	PRIME
79	6361	PRIME	98	9743	PRIME
80	6521	PRIME	99	9941	PRIME
81	6683	COMPOSITE	100	10141	PRIME
82	6847	COMPOSITE			

Euler's Formula

In 1963, Stanislaw Ulam, the brilliant mathematician who did pioneering work on the atomic bomb at Los Alamos, was doodling numbers on a slip of paper. He scribbled consecutive whole numbers, starting with 1, in a kind of square spiral radiating outward.

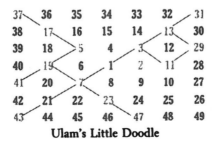

Ulam's Little Doodle

To his surprise, the prime numbers in his doodle, which I've marked in gray, tended to fall on diagonal lines. Inspired by this serendipitous discovery, Ulam and two coworkers, Mark Wells and Myron Stein, investigated square spirals that started with whole numbers other than 1. The whole numbers from 41 to 440 also form a spiral. Again, the primes often fall on diagonal

421	420	419	418	417	416	415	414	413	412	411	410	409	408	407	406	405	404	403	402
422	347	346	345	344	343	342	341	340	339	338	337	336	335	334	333	332	331	330	401
423	348	281	280	279	278	277	276	275	274	273	272	271	270	269	268	267	266	329	400
424	349	282	223	222	221	220	219	218	217	216	215	214	213	212	211	210	265	328	399
425	350	283	224	173	172	171	170	169	168	167	166	165	164	163	162	209	264	327	398
426	351	284	225	174	131	130	129	128	127	126	125	124	123	122	161	208	263	326	397
427	352	285	226	175	132	97	96	95	94	93	92	91	90	121	160	207	262	325	396
428	353	286	227	176	133	98	71	70	69	68	67	66	89	120	159	206	261	324	395
429	354	287	228	177	134	99	72	53	52	51	50	65	88	119	158	205	260	323	394
430	355	288	229	178	135	100	73	54	43	42	49	64	87	118	157	204	259	322	393
431	356	289	230	179	136	101	74	55	44	41	48	63	86	117	156	203	258	321	392
432	357	290	231	180	137	102	75	56	45	46	47	62	85	116	155	202	257	320	391
433	358	291	232	181	138	103	76	57	58	59	60	61	84	115	154	201	256	319	390
434	359	292	233	182	139	104	77	78	79	80	81	82	83	114	153	200	255	318	389
435	360	293	234	183	140	105	106	107	108	109	110	111	112	113	152	199	254	317	388
436	361	294	235	184	141	142	143	144	145	146	147	148	149	150	151	198	253	316	387
437	362	295	236	185	186	187	188	189	190	191	192	193	194	195	196	197	252	315	386
438	363	296	237	238	239	240	241	242	243	244	245	246	247	248	249	250	251	314	385
439	364	297	298	299	300	301	302	303	304	305	306	307	308	309	310	311	312	313	384
440	365	366	367	368	369	370	371	372	373	374	375	376	377	378	379	380	381	382	383

Ulam's Big Doodle

lines. The main diagonal running from 421 to 383 corresponds to the primes generated by Euler's formula $n^2 + n + 41$.

In 1963, the Maniac II mainframe at Los Alamos had in its memory the first 90 million primes. "At Los Alamos, we also had one of the first graphics facilities," recalls Wells, "so we got very excited about having the computer plot the primes." Maniac II drew a square-spiral diagram for all the primes under 10 million. Sure enough, many uncannily preferred diagonal lines.

Maniac's Doodles

Euler's formula $n^2 + n + 41$ turns out to be surprisingly good for large values of n. Maniac II computed that for the primes under 10 million, the formula generated primes 47.5 percent of the time. The formula works even better for lower values of n. For values of n under 2,398, there's an even chance of getting a prime. And for values of n under 100, the formula yields 86 primes and only 14 composite numbers.

Ulam and his coworkers discovered other prime-generating formulas that are almost as good as Euler's. With a success rate of 46.6 percent, the formula $4n^2 + 170n + 1,847$ yields 760 primes below 10 million that are not generated by Euler's formula. And the formula $4n^2 + 4n + 59$, with a success rate of 43.7 percent, yields some 1,500 primes not generated by the other two formulas.

The biggest paradox of all is that, despite these high success rates, despite the apparent diagonal-line regularity in the square spirals, number theorists have proved that no formula vaguely resembling Euler's can generate all the primes or nothing but primes. Nonetheless, this proof has not deterred romantics from looking for patterns to the primes.

Of the first 100 numbers, 25 are prime: 2, 3, 5, 7, 11, 13, 17, 19, 23, 29, 31, 37, 41, 43, 47, 53, 59, 61, 67, 71, 73, 79, 83, 89, and 97. The spacing between these consecutive primes (and the infinitely many others that succeed them) follows no obvious pattern. Since 2 is the only even prime number, 2 and 3 are the only prime numbers that can differ by one.

What about primes—called twin primes—that differ by two? Among the first 25 primes are eight pairs of twin primes: (3, 5), (5, 7), (11, 13), (17, 19), (29, 31), (41, 43), (59, 61), and (71, 73). For almost 150 years, number theorists have conjectured that pairs of twin primes are inexhaustible, like the primes themselves, but no one has been able to prove this. Progress was made in 1966, when the Chinese mathematician Chen Jing-run proved that there exist infinitely many pairs of numbers that differ by two in which the first number is a prime and the second is either a prime or the product of two primes. (A number that is the product of two primes is called "almost prime," a description that attests to the irrepressible optimism of mathematicians as well as to the intractability of bona fide prime numbers.)

In another display of optimism, Chen proved a weaker version of Goldbach's conjecture: every "sufficiently large" even number is the sum of a prime and an almost prime. "Sufficiently large" is a euphemism in the prime-number literature for "I know my proof is true for all numbers beyond some number Q, but I don't know what Q is." Despite the vagueness of the phrase "sufficiently large," mathematicians consider Chen's proof to be the most significant contribution to prime-number theory in the past three decades.

Much more is known about how far apart primes are than about how close together they are. Indeed, it is easy to prove that there are arbitrarily long sequences of consecutive numbers that are not prime. Let $n!$ represent the product of all the whole numbers from 1 to n. By its construction, $n!$ can be evenly divided by every whole number from 2 to n. Imagine the sequence of consecutive numbers $n! + 2$, $n! + 3$, $n! + 4$, and so on all the way to $n! + n$. Now, the first term in the sequence, $n! + 2$, is evenly divisible by 2; the second term, $n! + 3$, is evenly divisible by 3; the third term, $n! + 4$, is evenly divisible by 4; and so on. There are $n - 1$ numbers in this sequence, and none is a prime number. By choosing n to be as large as you want, you can have a prime-free sequence of consecutive whole numbers as long as you want.

Prime-rich sequences, however, are also abundant. In fact, number theorists believe that the prime numbers can form arbitrarily long arithmetic progressions (sequences of primes separated by a common difference). Short arithmetic progressions of primes are easy to find. For example, the primes 3, 5, and 7 form a progression of three terms with a common difference of 2. (In 1944, it was proven that there are infinitely many sets of three primes in arithmetic progression.) And the primes 199, 409, 619, 829, 1,039, 1,249, 1,459, 1,669, 1,879, and 2,089 constitute a progression of ten terms with a common difference of 210. For longer progressions, the initial prime and the common difference escalate rapidly, making them difficult to find. In 1983, however, Paul Pritchard at Cornell found 19 primes in arithmetic progression; the initial prime is 8,297,-644,387, and the common difference is 4,180,566,390.

Some mathematicians have even conjectured that there are arbitrarily long arithmetic progressions of *consecutive* primes. For example, the consecutive primes 1,741, 1,747, 1,753, and 1,759 form a four-term progression with a common difference of six. Nevertheless, at this point no one has been able to prove this conjecture, let alone the weaker conjecture about arithmetic progressions in which the primes need not be consecutive.

A mammoth treatise could be written about what is known and not known about primes. One more simple example will suffice. It has been proved that there is at least one prime between any number greater than 1 and its double. (A surprising consequence of this proof is that there are at least three primes having exactly n digits, where n is any positive integer whatever.) But no one knows whether a prime lies between the square of any number greater than 1 and the square of its successor.

Since the prime numbers themselves follow no known pattern, it is perhaps only fitting that no rhyme or reason is apparent in what mathematicians have been able to prove about them. Some basic theorems—that there are infinitely many primes, that there are arbitrarily wide gaps between them—have proofs that could not be simpler. Other theorems, such as Goldbach's conjecture, have resisted proof, although no self-respecting mathematician doubts their truth. To make progress, number theorists have resorted to proving theorems about "almost primes" and "sufficiently large numbers." What the field needs is another Euclid or Euler. Until then, we may remain in the curious situation where forces in government and industry that depend on secret communications continue to profit from the ignorance of mathematicians.

Readers who are taken by number theory can try their hand—and their calculators—at these unproven conjectures. If the conjectures are true, the proofs will presumably draw on technical mathematical results that are beyond the scope of laymen. But if—contrary to expectation—they happen to be false, all that's required is a counterexample. The best mathematical minds have been known to slip up. Euler claimed that a fifth power can never equal the sum of two fifth powers, three fifth powers, or four fifth powers. (In other words, there are no integers x, y, and z that satisfy the equation $x^5 = y^5 + z^5$, no integers a, b, c, and d that satisfy the equation $a^5 = b^5 + c^5 + d^5$, and no integers m, no, o, p and q that satisfy the equation $m^5 = n^5 + o^5 + p^5 + q^5$.) Two centuries later, in 1966, this claim was refuted by the discovery of a counterexample: 144 raised to the fifth power turns out to equal the sum of four fifth powers, namely, those of 27, 84, 110, and 133.

If contemplating unproved conjectures is not your thing, perhaps thinking about specific numbers is. But don't make Hardy's mistake of prematurely dismissing a taxicab number as uninteresting. I was recently on a long plane flight, absorbed in a novel, when the fidgeting fellow in the next seat made an awkward stab at initiating conversation: "We are on flight 407. That number seems dull to me, which I hope isn't a bad omen."

"Nonsense," I replied, without looking up from my reading. "The number isn't dull at all. It's quite interesting. It's the largest three-digit number that equals the sum of the cubes of its digits." The man looked at me as if I were crazy, but he took out a pad and started scribbling numbers. He did this for the duration of the flight, and I was able to finish my novel uninterrupted.

4

THE CRYPTIC CASE OF A SWARTHY STRANGER

Cryptography—the science of making and breaking codes—is an increasingly quantitative discipline, practiced by mathematicians who have access to the latest in computer technology. The ciphers used today in the military and in private industry barely resemble the ciphers of yesteryear and are generally much harder to break. And yet, despite these advances, in many situations the newfangled mathematical codes are not useful, and some age-old ciphers remain immune to state-of-the-art code-cracking techniques.

Cryptography certainly has come a long way since the first century B.C., when Julius Caesar reportedly used a naively simple substitution cipher, in which each letter was replaced by the letter that followed it alphabetically by three places. Caesar's confidants would have understood him had he said "Hw wx, Euxwh!" instead of "Et tu, Brute!" Remarkably, almost two thousand years later, the Confederate generals A. S. Johnson and Pierre Beauregard, resurrected this simple cipher during the Battle of Shiloh.

A cipher found in the Old Testament is just as simple. In Jeremiah (25:26 and 51:41), the prophet wrote *She-shach* for Babel. The second letter of the Hebrew alphabet *(b)* was replaced by the second-to-last letter *(sh)*, and the twelfth letter *(l)* was replaced by the twelfth-to-last letter *(ch)*. (The vowels are in the wrong order, but in Hebrew vowels are of secondary importance.) The cipher is called Athbash, an acronym formed from the first Hebrew letter *(a)*, the last letter *(th)*, the second letter *(b)*, and the second-to-last letter *(sh)*.

The drawback of an elementary substitution cipher is that it can be cracked simply by analyzing the frequency with which each symbol appears. In every language, the letters of a lengthy plaintext (the intelligible message) exhibit a predictable frequency. For example, *e* is the most common letter in English, showing up an eighth of the time. It is a good assumption that the most common symbol of a long ciphertext (the concealed message) represents the letter *e*. If, on the basis of a frequency count, the cryptanalyst can decipher the nine most common letters, *e, t, a, o, n, i, r, s,* and *h,* respectively, he has generally broken 70 percent of the cipher. The most modern of code-cracking techniques is based on the age-old method of frequency analysis.

Frequency analysis also applies to the positions of letters within words and to combinations of letters. For example, the first letter of more than half of all English words comes from the five-letter group *t, a, o, s,* and *w.* Only ten words *(the, of, and, to, a, in, that, it, is,* and *I)* make up more than a quarter of the typical English text.

The larger the number of words enciphered, the easier it is to crack the cipher by frequency analysis. In the heat of battle, there is generally no shortage of ciphertext, as messages are continually radioed back and forth between field stations and headquarters. In World War I, the Germans transmitted by radio two million enciphered words a month. In World War II, the Allied Supreme Headquarters often sent out more than two million enciphered words a day.

A cipher, like Caesar's or Athbash, in which the assignment of cipher symbols to plaintext follows a pattern, is particularly vunerable because the pattern can be discovered with minimal effort. For example, if frequency analysis of a sample of Caesar ciphertext suggests that *h* stands for *e*, *w* for *t*, and *d* for *a*, the cryptoanalyst will suspect that each cipher letter stands for the plaintext letter that precedes it alphabetically by three places. He will then check to see if his suspicion was correct. Hunches and guesses, of course, are central to code breaking, because it is easy to pursue them and see whether they pan out.

If it hadn't been for frequency analysis, Mary Queen of Scots might have kept her head. In the simple substitution cipher that she used to write her perfidious correspondence, she showed herself to be wiser than both Caesar and Jeremiah. She assigned the cipher symbols randomly, and she peppered her correspondence with "nulls," meaningless symbols.

a b c d e f g h i j k l m n o p q r s t v w x y z

nulls

Nevertheless, Sir Francis Walsingham, the founder of the British Secret Service, managed to weed out the nulls and do a frequency count of the remaining symbols. He decrypted Mary's plot to assassinate Queen Elizabeth and inherit the throne. On the basis of this cryptoanalysis, Mary was convicted of treason and executed.

If she had known of the work of the fifteenth-century Italian architect Leon Battista Alberti, she might have avoided the chopping block. To sabotage a frequency count, Alberti came up with an amazing scheme—"worthy of kings," he said—in which every plaintext letter could be represented by every cipher symbol. In essence, more than one cipher alphabet serves to encrypt a given message. Called a polyalphabetic cipher; Alberti's idea is the basis of modern cryptography.

Alberti's system makes use of the following table. Above the table are uppercase letters known as key letters, which identify the cipher alphabets in the table. To the left of the table are plaintext letters, also in upper case.

KEY LETTERS

	A	B	C	D	E	F	G	H	I	J	K	L	M	N	O	P	Q	R	S	T	U	V	W	X	Y	Z
A	a	b	c	d	e	f	g	h	i	j	k	l	m	n	o	p	q	r	s	t	u	v	w	x	y	z
B	b	c	d	e	f	g	h	i	j	k	l	m	n	o	p	q	r	s	t	u	v	w	x	y	z	a
C	c	d	e	f	g	h	i	j	k	l	m	n	o	p	q	r	s	t	u	v	w	x	y	z	a	b
D	d	e	f	g	h	i	j	k	l	m	n	o	p	q	r	s	t	u	v	w	x	y	z	a	b	c
E	e	f	g	h	i	j	k	l	m	n	o	p	q	r	s	t	u	v	w	x	y	z	a	b	c	d
F	f	g	h	i	j	k	l	m	n	o	p	q	r	s	t	u	v	w	x	y	z	a	b	c	d	e
G	g	h	i	j	k	l	m	n	o	p	q	r	s	t	u	v	w	x	y	z	a	b	c	d	e	f
H	h	i	j	k	l	m	n	o	p	q	r	s	t	u	v	w	x	y	z	a	b	c	d	e	f	g
I	i	j	k	l	m	n	o	p	q	r	s	t	u	v	w	x	y	z	a	b	c	d	e	f	g	h
J	j	k	l	m	n	o	p	q	r	s	t	u	v	w	x	y	z	a	b	c	d	e	f	g	h	i
K	k	l	m	n	o	p	q	r	s	t	u	v	w	x	y	z	a	b	c	d	e	f	g	h	i	j
L	l	m	n	o	p	q	r	s	t	u	v	w	x	y	z	a	b	c	d	e	f	g	h	i	j	k
M	m	n	o	p	q	r	s	t	u	v	w	x	y	z	a	b	c	d	e	f	g	h	i	j	k	l
N	n	o	p	q	r	s	t	u	v	w	x	y	z	a	b	c	d	e	f	g	h	i	j	k	l	m
O	o	p	q	r	s	t	u	v	w	x	y	z	a	b	c	d	e	f	g	h	i	j	k	l	m	n
P	p	q	r	s	t	u	v	w	x	y	z	a	b	c	d	e	f	g	h	i	j	k	l	m	n	o
Q	q	r	s	t	u	v	w	x	y	z	a	b	c	d	e	f	g	h	i	j	k	l	m	n	o	p
R	r	s	t	u	v	w	x	y	z	a	b	c	d	e	f	g	h	i	j	k	l	m	n	o	p	q
S	s	t	u	v	w	x	y	z	a	b	c	d	e	f	g	h	i	j	k	l	m	n	o	p	q	r
T	t	u	v	w	x	y	z	a	b	c	d	e	f	g	h	i	j	k	l	m	n	o	p	q	r	s
U	u	v	w	x	y	z	a	b	c	d	e	f	g	h	i	j	k	l	m	n	o	p	q	r	s	t
V	v	w	x	y	z	a	b	c	d	e	f	g	h	i	j	k	l	m	n	o	p	q	r	s	t	u
W	w	x	y	z	a	b	c	d	e	f	g	h	i	j	k	l	m	n	o	p	q	r	s	t	u	v
X	x	y	z	a	b	c	d	e	f	g	h	i	j	k	l	m	n	o	p	q	r	s	t	u	v	w
Y	y	z	a	b	c	d	e	f	g	h	i	j	k	l	m	n	o	p	q	r	s	t	u	v	w	x
Z	z	a	b	c	d	e	f	g	h	i	j	k	l	m	n	o	p	q	r	s	t	u	v	w	x	y

(PLAINTEXT is printed vertically along the left margin of the table.)

Before any messages can be sent, the communicating parties must agree on a kind of password called a keyword. To encrypt a message, the keyword is written repeatedly above the plain-

text. Say the keyword is LOVE and the plaintext message is
SEND MORE MONEY. The sender would write,

keyword: L O V E L O V E L O V E L
plaintext: S E N D M O R E M O N E Y

The keyword letter above each plaintext letter indicates
which cipher alphabet in the table should be used for enciphering that particular plaintext letter. The S in SEND should be
represented by the L alphabet (because the L in LOVE falls
above the S in SEND), and so the ciphertext letter, d, is found
in the table at the intersection of the S row and the L column:

KEY LETTERS

```
      A B C D E F G H I J K L M N O P Q R S T U V W X Y Z
   A  a b c d e f g h i j k l m n o p q r s t u v w x y z
   B  b c d e f g h i j k l m n o p q r s t u v w x y z a
   C  c d e f g h i j k l m n o p q r s t u v w x y z a b
   D  d e f g h i j k l m n o p q r s t u v w x y z a b c
   E  e f g h i j k l m n o p q r s t u v w x y z a b c d
   F  f g h i j k l m n o p q r s t u v w x y z a b c d e
   G  g h i j k l m n o p q r s t u v w x y z a b c d e f
   H  h i j k l m n o p q r s t u v w x y z a b c d e f g
   I  i j k l m n o p q r s t u v w x y z a b c d e f g h
   J  j k l m n o p q r s t u v w x y z a b c d e f g h i
   K  k l m n o p q r s t u v w x y z a b c d e f g h i j
   L  l m n o p q r s t u v w x y z a b c d e f g h i j k
   M  m n o p q r s t u v w x y z a b c d e f g h i j k l
   N  n o p q r s t u v w x y z a b c d e f g h i j k l m
   O  o p q r s t u v w x y z a b c d e f g h i j k l m n
   P  p q r s t u v w x y z a b c d e f g h i j k l m n o
   Q  q r s t u v w x y z a b c d e f g h i j k l m n o p
   R  r s t u v w x y z a b c d e f g h i j k l m n o p q
   S  s t u v w x y z a b c d e f g h i j k l m n o p q r
   T  t u v w x y z a b c d e f g h i j k l m n o p q r s
   U  u v w x y z a b c d e f g h i j k l m n o p q r s t
   V  v w x y z a b c d e f g h i j k l m n o p q r s t u
   W  w x y z a b c d e f g h i j k l m n o p q r s t u v
   X  x y z a b c d e f g h i j k l m n o p q r s t u v w
   Y  y z a b c d e f g h i j k l m n o p q r s t u v w x
   Z  z a b c d e f g h i j k l m n o p q r s t u v w x y
```

PLAINTEXT

Likewise, the *E* in SEND is represented by the *O* alphabet, and so the ciphertext symbol, *s*, is at the intersection of the *E* row and the *O* column:

KEY LETTERS

	A	B	C	D	E	F	G	H	I	J	K	L	M	N	O	P	Q	R	S	T	U	V	W	X	Y	Z
A	a	b	c	d	e	f	g	h	i	j	k	l	m	n	o	p	q	r	s	t	u	v	w	x	y	z
B	b	c	d	e	f	g	h	i	j	k	l	m	n	o	p	q	r	s	t	u	v	w	x	y	z	a
C	c	d	e	f	g	h	i	j	k	l	m	n	o	p	q	r	s	t	u	v	w	x	y	z	a	b
D	d	e	f	g	h	i	j	k	l	m	n	o	p	q	r	s	t	u	v	w	x	y	z	a	b	c
E	e	f	g	h	i	j	k	l	m	n	o	p	q	r	s	t	u	v	w	x	y	z	a	b	c	d
F	f	g	h	i	j	k	l	m	n	o	p	q	r	s	t	u	v	w	x	y	z	a	b	c	d	e
G	g	h	i	j	k	l	m	n	o	p	q	r	s	t	u	v	w	x	y	z	a	b	c	d	e	f
H	h	i	j	k	l	m	n	o	p	q	r	s	t	u	v	w	x	y	z	a	b	c	d	e	f	g
I	i	j	k	l	m	n	o	p	q	r	s	t	u	v	w	x	y	z	a	b	c	d	e	f	g	h
J	j	k	l	m	n	o	p	q	r	s	t	u	v	w	x	y	z	a	b	c	d	e	f	g	h	i
K	k	l	m	n	o	p	q	r	s	t	u	v	w	x	y	z	a	b	c	d	e	f	g	h	i	j
L	l	m	n	o	p	q	r	s	t	u	v	w	x	y	z	a	b	c	d	e	f	g	h	i	j	k
M	m	n	o	p	q	r	s	t	u	v	w	x	y	z	a	b	c	d	e	f	g	h	i	j	k	l
N	n	o	p	q	r	s	t	u	v	w	x	y	z	a	b	c	d	e	f	g	h	i	j	k	l	m
O	o	p	q	r	s	t	u	v	w	x	y	z	a	b	c	d	e	f	g	h	i	j	k	l	m	n
P	p	q	r	s	t	u	v	w	x	y	z	a	b	c	d	e	f	g	h	i	j	k	l	m	n	o
Q	q	r	s	t	u	v	w	x	y	z	a	b	c	d	e	f	g	h	i	j	k	l	m	n	o	p
R	r	s	t	u	v	w	x	y	z	a	b	c	d	e	f	g	h	i	j	k	l	m	n	o	p	q
S	s	t	u	v	w	x	y	z	a	b	c	d	e	f	g	h	i	j	k	l	m	n	o	p	q	r
T	t	u	v	w	x	y	z	a	b	c	d	e	f	g	h	i	j	k	l	m	n	o	p	q	r	s
U	u	v	w	x	y	z	a	b	c	d	e	f	g	h	i	j	k	l	m	n	o	p	q	r	s	t
V	v	w	x	y	z	a	b	c	d	e	f	g	h	i	j	k	l	m	n	o	p	q	r	s	t	u
W	w	x	y	z	a	b	c	d	e	f	g	h	i	j	k	l	m	n	o	p	q	r	s	t	u	v
X	x	y	z	a	b	c	d	e	f	g	h	i	j	k	l	m	n	o	p	q	r	s	t	u	v	w
Y	y	z	a	b	c	d	e	f	g	h	i	j	k	l	m	n	o	p	q	r	s	t	u	v	w	x
Z	z	a	b	c	d	e	f	g	h	i	j	k	l	m	n	o	p	q	r	s	t	u	v	w	x	y

PLAINTEXT

Carrying out this procedure for the entire message, we find that SEND MORE MONEY produces the ciphertext DSIHXCMIXCIIJ:

keyword: L O V E L O V E L O V E L
plaintext: S E N D M O R E M O N E Y
ciphertext: D S I H X C M I X C I I J

Decryption is a similar process: the keyword is written repeatedly above the ciphertext, and the plaintext is extracted from the appropriate alphabets in the table. The beauty of the system is that even if an eavesdropper got hold of the table, he would not get very far without the keyword. In wartime, the keywords are frequently changed, for extra security.

But careless use of the best cipher can compromise its security, making code breaking much easier in practice than in theory. Diplomatic and military communications often begin and end with characteristic pleasantries ("Greetings!" and "Respectfully yours"), which are footholds for the cryptoanalyst. Certain proper names—especially ones that are unusually long—can also give the show away. For example, in World War II, German communications spoke in cipher of the *Wehrmachtnachrichtenverbindungen,* the Communications Intelligence Service of the Germany Army.

Information can often be coaxed out of the enemy. In May 1942, the American high command knew that a vast Japanese force of eleven battleships, five carriers, sixteen cruisers, and forty-nine destroyers was going to strike soon, but it did not know where. Japanese radio dispatchers referred again and again to AF. Did AF stand for California, Alaska, Midway Island, or some other place? To find out, American intelligence agents instructed the U.S. garrison at Midway to radio Pearl Harbor that it was running out of water. The garrison complied. Shortly thereafter, the Americans intercepted a Japanese dispatch that reported a water shortage at AF. When the attack came, the Americans were prepared. With a numerically inferior force, they repelled the Japanese and won the great naval battle of Midway.

Even if a cipher is not compromised, it may be broken because it has intrinsic weaknesses overlooked by the sender but exploited by the wily eavesdropping cryptoanalyst. For three hundred years Alberti's polyalphabetic cipher was thought to be invulnerable, but then, in the 1860s, Friedrich W. Kasiska, a former Prussian infantryman, discovered a few built-in weaknesses. He found, for example, that if a sequence of plaintext letters that comes up more than once happens to be enciphered each time by the same keyword letters, identical ciphertext results. For example, in the message SEND MORE MONEY, the key-letter sequence *LO* twice enciphers the plaintext sequence *MO* as *XC*:

keyword: L O V E L O V E L O V E L
plaintext: S E N D M O R E M O N E Y
ciphertext: D S I H X C M I X C I I J

The repeated ciphertext XC indicates the length of the keyword. In general, the number of ciphertext letters from the start of one instance of repeated text to the start of another instance is a multiple of the number of letters in the keyword. If bits of the ciphertext repeat often enough, the cryptoanalyst can figure out the length of the keyword and, hence, the number of cipher alphabets employed. Then it's just a matter of cataloging which ciphertext letters came from which cipher alphabet. For each cipher alphabet, a frequency count will reveal the plaintext letters.

In Alberti's cipher, the method of encryption—the table of cipher alphabets—can be made public, so long as the keyword is kept secret, without jeopardizing the security of the cipher. As we saw in the last chapter, modern cryptographers, drawing on innovative mathematical methods, have been able to take this trend to an amazing extreme: both the encryption method and the key itself can be made public without compromising the cipher. In other words, the power to encipher a message is not the same as the power to decipher it.

In this day and age, when cryptography is increasingly computerized, breakdowns in technology can have severe consequences. If a situation ever required secret communication, using one of the virtually impenetrable ciphers that modern mathematics has to offer, it was in October 1985. Early one morning, the Reagan administration learned from intelligence sources that President Hosni Mubarak of Egypt was lying about the whereabouts of the four Palestinian terrorists who had hijacked the Italian cruise ship *Achille Lauro* and murdered sixty-nine-year-old Leon Klinghoffer in his wheelchair. Contrary to what Mubarak had publicly stated, the hijackers were still on Egyptian soil, preparing to leave the country quietly by air. While U.S. intelligence sources managed to locate the plane that the terrorists planned to take—an Egypt-Air Boeing 737 jetliner sitting on the runway of an air base near Cairo—the Pentagon's counterterrorist experts hurriedly came up with a plan for intercepting the civilian getaway plane with F-14 Tomcats, backed by reconnaissance and radar-jamming aircraft.

Meanwhile, for President Reagan, it was business as usual. As the CIA and the Pentagon went into overdrive, Reagan lunched on baked goods in the Sara Lee kitchens outside Chicago. Postprandial chitchat was interrupted by an urgent (and secret) communication from Washington; the president's advisers outlined for him their audacious plan to force down the EgyptAir jet. Reagan liked what he heard but before giving a firm go-ahead wanted to know how many lives would be at risk. A few hours later, en route to Washington aboard *Air Force One,* Reagan called Defense Secretary Caspar Weinberger, who was flying on a military plane to his summer home in Bar

Harbor, Maine. On an open shortwave radio channel, the president—speaking as he normally does, not in code, and with his voice not scrambled by high-tech gadgetry—ordered the reluctant Weinberger to proceed with the daring mission. An amateur radio operator overheard every word of the president's provocative order, and the brother of the eavesdropper lost no time in contacting CBS News, which chose not to report the president's order. Not being "Page Six" of the *New York Post,* CBS News wanted a firsthand source, either the radio operator himself, not his brother, or, better yet, a tape recording of the overheard conversation. A few hours later the EgyptAir jet was forced down, in exactly the manner told to CBS News.

The *New York Times* subsequently quoted a White House official as saying, "They [Reagan and Weinberger] were on two different planes with two different cryptographic systems. They could have been patched through another cryptographic system, but time was of the essence and they decided to go nonsecure. They felt that the information was not sensitive enough that they needed a secure call." But you can be sure that if an amateur radio operator overheard their conversation, the Soviet Union, which apparently monitors all radio transmissions from *Air Force One,* overheard it too. If the Kremlin had not shown restraint, the American F-14 fighters might have encountered a squadron of Soviet MIGs instead of a defenseless civilian jetliner.

When time is of the essence, codes that require elaborate machinery or sophisticated mathematical methods are impractical. In the heat of battle, for example, orders must be acted on as soon as they are received; there's not much time for deciphering. If Reagan and Weinberger had known a relatively obscure foreign tongue, they could have spoken it. In the Boer War, the British runners who delivered messages between encampments spoke Latin. That, at least, put a tiny obstacle in the way of eavesdroppers.

In World War I, the commander of the American Expeditionary Force in France, fearing that the Germans were listening in on every communication, discovered a unique communications resource in the American Indians in his regiment who spoke a total of twenty-six recondite languages, only five of which had written characters. As eight Choctaw Indians spread the word by field telephone, he orchestrated "a delicate withdrawal" of the Second Battalion from Chuffilly. Before the United States entered World War II, the military studied many native American languages and identified Navaho as ideal for battlefield communication.

The language was apparently known by only twenty-eight people outside the tribe, none of them affiliated with the enemy. Navaho, like Chinese, is extremely difficult to learn, because the meaning of words turns on subtle variations in pronunciation. Moreover, there was no Berlitz crash course in Navaho; it could be learned only from native speakers, all of whom, fortunately for the Pentagon, were here in the United States. And since the Navahos numbered more than 50,000, there were surely many able-bodied men who could be conscripted. By the end of the war, 420 Navahos had helped the Marines advance from the Solomon Islands to Okinawa, where they were particularly helpful, barking orders in a peculiar language that left the Japanese high command totally baffled but made for swiftly executed maneuvers.

Despite the increasing number of mathematicians drawn to cryptography, and the increasing supercomputer resources harnessed for code making and breaking, old ciphers still confound—and distract—the experts. Written more than a century and half ago, the notorious Beale ciphers, which apparently conceal the whereabouts of a $17 million treasure trove, still occupy the efforts of "at least 10 percent of the best cryptoanalytic minds in the country," says Carl Hammer, former chief computer scientist at Sperry Univac for two decades and a pioneer in the computer analysis of the statistical properties of ciphers. "And not a dime of this effort should be begrudged," Hammer adds. "The work—even the lines that have led into blind alleys—has more than paid for itself in advancing and refining computer research."

The legacy of the Beale ciphers goes back to January 1820, when a tall, swarthy, ruggedly handsome stranger with jet black eyes and jet black hair, worn longer than the style of the day dictated, arrived on horseback at the Washington Hotel in Lynchburg, Virginia. Greeted by Robert Morriss, the hotel's proprietor, who was known to the rich for his conviviality and to the poor for his generosity, the stranger introduced himself as Thomas Jefferson Beale. After touring the premises and inspecting the accommodations offered him and his horse, Beale told Morriss that he planned to stay for the winter. A spirited conversationalist, Beale proved to be an affable guest, his manly beauty favored by the ladies and envied by the men. He regaled the other guests with long stories on every conceivable subject except that of his family, his lineage, and his residence—of which he never said anything. Late that March, he quietly departed for an undisclosed destination.

For nearly two years, no one heard from him. Then, in January 1822, he showed up unannounced at the hotel, the same genial man as before, only swarthier and handsomer than ever, his trim and tan body suggesting he had been on an extraordinary outdoor adventure. Everyone, particularly the women, welcomed his return. Come spring, Beale again disappeared, leaving behind a locked iron box that Morriss was supposed to keep safe until his return. That summer Morriss received a letter from Beale, with the dateline St. Louis, May 9, in which he described encountering buffalo and savage grizzlies. (Back then, St. Louis was a tiny frontier town.) "How long I may be absent I cannot now determine," the letter continued, "certainly not less than two years, perhaps longer.

"With regard to the box left in your charge I have few words to say. . . . It contains papers vitally affecting the fortunes of myself and many others engaged in business with me, and in the event of my death its loss might be irreparable. You will, therefore, see the necessity of guarding it with vigilance and care to prevent so great a catastrophe. It also contains some letters addressed to yourself and which will be necessary to enlighten you concerning the business in which we are engaged. . . . [Should I or my associates not claim the box within ten years from the date of this letter], you will open it, which can be done by removing the lock.

"You will find, in addition to the papers addressed to you, other papers which will be unintelligible without the aid of a key to assist you. Such a key I have left in the hands of a friend in this place, sealed, addressed to yourself, and endorsed 'Not to be delivered until June 1832.' By means of this you will understand fully all you will be required to do. . . . With kindest wishes for your most excellent wife, compliments to the ladies, a good word to enquiring friends, if there be any, and assurances of my highest esteem for yourself, I remain, as ever, Your sincere friend, Tho. Jeff. Beale."

Needless to say, Morriss never heard from Beale again. Whether he was massacred by Indians, or mutilated by savage animals, or whether exposure, and perhaps privation, took its toll is only a matter for speculation. The summer of 1832 came around, and Morriss did not receive the promised key from St.

Louis. By authority of Beale's letter, Morriss could have broken the box open that year, but, busy with other duties, he waited until 1845. Inside he found two letters addressed to him—a long, informative one and a short, uninteresting one—some old receipts, and a few sheets of paper covered with strings of numbers.

The long letter, dated January 4, 1822, began, "You will, doubtless, be surprised when you discover, from a perusal of this letter, the importance of the trust confided to you, the confidence reposed in your honor, by parties whom you have never seen and whose names you have never heard. The reasons are simple and easily told. It was imperative upon us that some one here should be selected to carry out our wishes in case of accident to ourselves, and your reputation as a man of integrity, unblemished honor, and business sagacity, influenced them to select you in place of others better known but, perhaps, not so reliable as yourself. It was with this design that I first visited your house, two years since, that I might judge by personal observation if your reputation was merited."

The letter went on to describe how Beale and a merry band of twenty-nine friends, all "fond of adventure, and if mixed with a little danger all the more acceptable," had embarked in April 1817 on a two-year hunting expedition to the far reaches of the great western plains. In the spring of 1818, some three hundred miles north of Sante Fe, the hunting party, suffering from fatigue, boredom, and inclement weather, followed an immense herd of buffalo into a deep ravine. Tired from the chase, the adventurers tethered their horses and set up camp. As they prepared the evening meal, a keen observer among them spied gold in a cleft in the rocks.

For the next eighteen months, the letter said, they mined gold, and silver too, having procured the assistance of friendly

Indians. Beale and a few of his companions then hauled the booty to Virginia, where they planned to stash it in a cave they had visited before, "near Buford's Tavern, in the county of Bedford." On reaching the cave, however, Beale found it unsatisfactory as a safe depository; "it was too frequently visited by the neighboring farmers, who used it as a receptacle for their sweet potatoes and other vegetables." And so he selected another hiding place.

Then he checked himself into the Washington Hotel. Satisfied that Morriss was as trustworthy as reputed, Beale ventured west again to rejoin his fellow miners. In the fall of 1822, he returned to Virginia, with large quantities of gold and silver, deposited the precious metals in the hiding place, and entrusted the locked box to Morriss.

As for the three unintelligible papers, which consisted entirely of numbers, the letter claimed that, when deciphered with the promised key, the papers would reveal the exact location of the hideaway, the precise contents of the stash, and the names and addresses of the thirty adventurers. The letter instructed Morriss to divide the treasure into thirty-one equal parts, retain one part for himself, as remuneration for his services, and distribute the other parts to the kin of the thirty claimants. "In conclusion, my dear friend," Beale wrote, "I beg that you will not allow any false or idle punctilio to prevent your receiving and appropriating the portion assigned to yourself. It is a gift, not from myself alone but from each member of our party, and will not be out of proportion to the services required of you."

Morriss's curiosity was undoubtedly piqued by the contents of the box. But he was motivated less by avarice than by the desire not to betray the confidence placed in him by the charismatic lady-killer and his twenty-nine unknown companions who were united by love of daring adventure and "the wild and roving character of their lives, the charms of which lured them farther and farther from civilization, until their lives were sacrificed to their temerity." Morriss devoted the remaining nineteen years of his life to recovering the treasure, but without the key to the mysterious papers, he was unable to make headway. Before he died, in 1863, he shared the contents of the box with James Ward, a bartender and family man of discretion who had accumulated sufficient savings to be able to spend his days searching for the elusive treasure.

Morriss thought that he was doing Ward a favor, potentially a very lucrative one, by letting him in on Beale's secret. Instead, it proved to be Ward's ruin. He became obsessed with the ciphers—all the more so since he managed to decode the second paper, which revealed the extent of the hidden treasure (2,921 pounds of gold, 5,100 pounds of silver, and, by today's standards, some $3.35 million worth of jewels) but not the burial site.

"It would be difficult to portray the delight he experienced," Ward wrote of himself, "when accident revealed to him the explanation [of the second paper]. But this accident, affording so much pleasure at the time, was a most unfortunate one for him, as it induced him to neglect family, friends, and all legitimate pursuits for what has proved, so far, the veriest illusion. . . . When the writer recalls his anxious hours, his midnight vigils, his toll, his hopes, and disappointments, all consequent upon this promise, he can only conclude that the legacy of Mr. Morriss was not as he designed it—a blessing in disguise."

It is time to take a look at the ciphers themselves:

Paper Number One: The Location of the $17 Million Treasure

71, 194, 38, 1701, 89, 76, 11, 83, 1629, 48, 94, 63, 132, 16, 111,
95, 84, 341, 975, 14, 40, 64, 27, 81, 139, 213, 63, 90, 1120, 8,
15, 3, 126, 2018, 40, 74, 758, 485, 604, 230, 436, 664, 582, 150,
251, 284, 308, 231, 124, 211, 486, 225, 401, 370, 11, 101, 305,
139, 189, 17, 33, 88, 208, 193, 145, 1, 94, 73, 416, 918, 263, 28,
500, 538, 356, 117, 136, 219, 27, 176, 130, 10, 460, 25, 485, 18,
436, 65, 84, 200, 283, 118, 320, 138, 36, 416, 280, 15, 71, 224,
961, 44, 16, 401, 39, 88, 61, 304, 12, 21, 24, 283, 134, 92, 63,
246, 486, 682, 7, 219, 184, 360, 780, 18, 64, 463, 474, 131, 160,
79, 73, 440, 95, 18, 64, 581, 34, 69, 128, 367, 460, 17, 81, 12,
103, 820, 62, 116, 97, 103, 862, 70, 60, 1317, 471, 540, 208,
121, 890, 346, 36, 150, 59, 568, 614, 13, 120, 63, 219, 812,
2160, 1780, 99, 35, 18, 21, 136, 872, 15, 28, 170, 88, 4, 30, 44,
112, 18, 147, 436, 195, 320, 37, 122, 113, 6, 140, 8, 120, 305,
42, 58, 461, 44, 106, 301, 13, 408, 680, 93, 86, 116, 530, 82,
568, 9, 102, 38, 416, 89, 71, 216, 728, 965, 818, 2, 38, 121, 195,
14, 326, 148, 234, 18, 55, 131, 234, 361, 824, 5, 81, 623, 48,
961, 19, 26, 33, 10, 1101, 365, 92, 88, 181, 275, 346, 201, 206,
86, 36, 219, 320, 829, 840, 68, 326, 19, 48, 122, 85, 216, 284,
919, 861, 326, 985, 233, 64, 68, 232, 431, 960, 50, 29, 81, 216,
321, 603, 14, 612, 81, 360, 36, 51, 62, 194, 78, 60, 200, 314,
676, 112, 4, 28, 18, 61, 136, 247, 819, 921, 1060, 464, 895, 10,
6, 66, 119, 38, 41, 49, 602, 423, 962, 302, 294, 875, 78, 14, 23,
111, 109, 62, 31, 501, 823, 216, 280, 34, 24, 150, 1000, 162,
286, 19, 21, 17, 340, 19, 242, 31, 86, 234, 140, 607, 115, 33,
191, 67, 104, 86, 52, 88, 16, 80, 121, 67, 95, 122, 216, 548, 96,
11, 201, 77, 364, 218, 65, 667, 890, 236, 154, 211, 10, 98, 34,
119, 56, 216, 119, 71, 218, 1164, 1496, 1817, 51, 39, 210, 36, 3,
19, 540, 232, 22, 141, 617, 84, 290, 80, 46, 207, 411, 150, 29,
38, 46, 172, 85, 194, 36, 261, 543, 897, 624, 18, 212, 416, 127,
931, 19, 4, 63, 96, 12, 101, 418, 16, 140, 230, 460, 538, 19, 27,
88, 612, 1431, 90, 716, 275, 74, 83, 11, 426, 89, 72, 84, 1300,
1706, 814, 221, 132, 40, 102, 34, 858, 975, 1101, 84, 16, 79, 23,
16, 81, 122, 324, 403, 912, 227, 936, 447, 55, 86, 34, 43, 212,
107, 96, 314, 264, 1065, 323, 328, 601, 203, 124, 95, 216, 814,

2906, 654, 820, 2, 301, 112, 176, 213, 71, 87, 96, 202, 35, 10, 2,
41, 17, 84, 221, 736, 820, 214, 11, 60, 760.

Paper Number Two: The Exact Nature of the Treasure

115, 73, 24, 818, 37, 52, 49, 17, 31, 62, 657, 22, 7, 15, 140, 47,
29, 107, 79, 84, 56, 238, 10, 26, 822, 5, 195, 308, 85, 52, 159,
136, 59, 210, 36, 9, 46, 316, 543, 122, 106, 95, 53, 58, 2, 42, 7,
35, 122, 53, 31, 82, 77, 250, 105, 56, 96, 118, 71, 140, 287, 28,
353, 37, 994, 65, 147, 818, 24, 3, 8, 12, 47, 43, 59, 818, 45, 316,
101, 41, 78, 154, 994, 122, 138, 190, 16, 77, 49, 102, 57, 72, 34,
73, 85, 35, 371, 59, 195, 81, 92, 190, 106, 273, 60, 394, 629,
270, 219, 106, 388, 287, 63, 3, 6, 190, 122, 43, 233, 400, 106,
290, 314, 47, 48, 81, 96, 26, 115, 92, 157, 190, 110, 77, 85, 196,
46, 10, 113, 140, 353, 48, 120, 106, 2, 616, 61, 420, 822, 29,
125, 14, 20, 37, 105, 28, 248, 16, 158, 7, 35, 19, 301, 125, 110,
496, 287, 98, 117, 520, 62, 51, 219, 37, 37, 113, 140, 818, 138,
549, 8, 44, 287, 388, 117, 18, 79, 344, 34, 20, 59, 520, 557, 107,
612, 219, 37, 66, 154, 41, 20, 50, 6, 584, 122, 154, 248, 110, 61,
52, 33, 30, 5, 38, 8, 14, 84, 57, 549, 216, 115, 71, 29, 85, 63, 43,
131, 29, 138, 47, 73, 238, 549, 52, 53, 79, 118, 51, 44, 63, 195,
12, 238, 112, 3, 49, 79, 353, 105, 56, 371, 566, 210, 515, 125,
360, 133, 143, 101, 15, 284, 549, 252, 14, 204, 140, 344, 26,
822, 138, 115, 48, 73, 34, 204, 316, 616, 63, 219, 7, 52, 150, 44,
52, 16, 40, 37, 157, 818, 37, 121, 12, 95, 10, 15, 35, 12, 131, 62,
115, 102, 818, 49, 53, 135, 138, 30, 31, 62, 67, 41, 85, 63, 10,
106, 818, 138, 8, 113, 20, 32, 33, 37, 353, 287, 140, 47, 85, 50,
37, 49, 47, 64, 6, 7, 71, 33, 4, 43, 47, 63, 1, 27, 609, 207, 229,
15, 190, 246, 85, 94, 520, 2, 270, 20, 39, 7, 33, 44, 22, 40, 7, 10,
3, 822, 106, 44, 496, 229, 353, 210, 199, 31, 10, 38, 140, 297,
61, 612, 320, 302, 676, 287, 2, 44, 33, 32, 520, 557, 10, 6, 250,
566, 246, 53, 37, 52, 83, 47, 320, 38, 33, 818, 7, 44, 30, 31, 250,
10, 15, 35, 106, 159, 113, 31, 102, 406, 229, 540, 320, 29, 66,
33, 101, 818, 138, 301, 316, 353, 320, 219, 37, 52, 28, 549, 320,
33, 8, 48, 107, 50, 822, 7, 2, 113, 73, 16, 125, 11, 110, 67, 102,
818, 33, 59, 81, 157, 38, 43, 590, 138, 19, 85, 400, 38, 43, 77,
14, 27, 8, 47, 138, 63, 140, 44, 35, 22, 176, 106, 250, 314, 216,

2, 10, 7, 994, 4, 20, 25, 44, 48, 7, 26, 46, 110, 229, 818, 190, 34, 112, 147, 44, 110, 121, 125, 96, 41, 51, 50, 140, 56, 47, 152, 549, 63, 818, 28, 42, 250, 138, 591, 98, 653, 32, 107, 140, 112, 26, 85, 138, 549, 50, 20, 125, 371, 38, 36, 10, 52, 118, 136, 102, 420, 150, 112, 71, 14, 20, 7, 24, 18, 12, 818, 37, 67, 110, 62, 33, 21, 95, 219, 520, 102, 822, 30, 38, 84, 305, 629, 15, 2, 10, 8, 219, 106, 353, 105, 106, 60, 242, 72, 8, 50, 204, 184, 112, 125, 549, 65, 106, 818, 190, 96, 110, 16, 73, 33, 818, 150, 409, 400, 50, 154, 285, 96, 106, 316, 270, 204, 101, 822, 400, 8, 44, 37, 52, 40, 240, 34, 204, 38, 16, 46, 47, 85, 24, 44, 15, 64, 73, 138, 818, 85, 78, 110, 33, 420, 515, 53, 37, 38, 22, 31, 10, 110, 106, 101, 140, 15, 38, 3, 5, 44, 7, 98, 287, 135, 150, 96, 33, 84, 125, 818, 190, 96, 520, 118, 459, 370, 653, 466, 106, 41, 107, 612, 219, 275, 30, 150, 105, 49, 53, 287, 250, 207, 134, 7, 53, 12, 47, 85, 63, 138, 110, 21, 112, 140, 495, 496, 515, 14, 73, 85, 584, 994, 150, 199, 16, 42, 5, 4, 25, 42, 8, 16, 822, 125, 159, 32, 204, 612, 818, 81, 95, 405, 41, 609, 136, 14, 20, 28, 26, 353, 302, 246, 8, 131, 159, 140, 84, 440, 42, 16, 822, 40, 67, 101, 102, 193, 138, 204, 51, 63, 240, 549, 122, 8, 10, 63, 140, 47, 48, 140, 288.

Paper Number Three: Names and Addresses of the Kin of the Adventurers

317, 8, 92, 73, 112, 89, 67, 318, 28, 96, 107, 41, 631, 78, 146, 397, 118, 98, 114, 246, 348, 116, 74, 88, 12, 65, 32, 14, 81, 19, 76, 121, 216, 85, 33, 66, 15, 108, 68, 77, 43, 24, 122, 96, 117, 36, 211, 301, 15, 44, 11, 46, 89, 18, 136, 68, 317, 28, 90, 82, 304, 71, 43, 221, 198, 176, 310, 319, 81, 99, 264, 380, 56, 37, 319, 2, 44, 53, 28, 44, 75, 98, 102, 37, 85, 107, 117, 64, 88, 136, 48, 151, 99, 175, 89, 315, 326, 78, 96, 214, 218, 311, 43, 89, 51, 90, 75, 128, 96, 33, 28, 103, 84, 65, 26, 41, 246, 84, 270, 98, 116, 32, 59, 74, 66, 69, 240, 15, 8, 121, 20, 77, 89, 31, 11, 106, 81, 191, 224, 328, 18, 75, 52, 82, 117, 201, 39, 23, 217, 27, 21, 84, 35, 54, 109, 128, 49, 77, 88, 1, 81, 217, 64, 55, 83, 116, 251, 269, 311, 96, 54, 32, 120, 18, 132, 102, 219, 211, 84, 150, 219, 275, 312, 64, 10, 106, 87, 75, 47, 21, 29, 37, 81, 44, 18, 126,

115, 132, 160, 181, 203, 76, 81, 299, 314, 337, 351, 96, 11, 28, 97, 318, 238, 106, 24, 93, 3, 19, 17, 26, 60, 73, 88, 14, 126, 138, 234, 286, 297, 321, 365, 264, 19, 22, 84, 56, 107, 98, 123, 111, 214, 136, 7, 33, 45, 40, 13, 28, 46, 42, 107, 196, 227, 344, 198, 203, 247, 116, 19, 8, 212, 230, 31, 6, 328, 65, 48, 52, 59, 41, 122, 33, 117, 11, 18, 25, 71, 36, 45, 83, 76, 89, 92, 31, 65, 70, 83, 96, 27, 33, 44, 50, 61, 24, 112, 136, 149, 176, 180, 194, 143, 171, 205, 296, 87, 12, 44, 51, 89, 98, 34, 41, 208, 173, 66, 9, 35, 16, 95, 8, 113, 175, 90, 56, 203, 19, 177, 183, 206, 157, 200, 218, 260, 291, 305, 618, 951, 320, 18, 124, 78, 65, 19, 32, 124, 48, 53, 57, 84, 96, 207, 244, 66, 82, 119, 71, 11, 86, 77, 213, 54, 82, 316, 245, 303, 86, 97, 106, 212, 18, 37, 15, 81, 89, 16, 7, 81, 39, 96, 14, 43, 216, 118, 29, 55, 109, 136, 172, 213, 64, 8, 227, 304, 611, 221, 364, 819, 375, 128, 296, 11, 18, 53, 76, 10, 15, 23, 19, 71, 84, 120, 134, 66, 73, 89, 96, 230, 48, 77, 26, 101, 127, 936, 218, 439, 178, 171, 61, 226, 313, 215, 102, 18, 167, 262, 114, 218, 66, 59, 48, 27, 19, 13, 82, 48, 162, 119, 34, 127, 139, 34, 128, 129, 74, 63, 120, 11, 54, 61, 73, 92, 180, 66, 75, 101, 124, 265, 89, 96, 126, 274, 896, 917, 434, 461, 235, 890, 312, 413, 328, 381, 96, 105, 217, 66, 118, 22, 77, 64, 42, 12, 7, 55, 24, 83, 67, 97, 109, 121, 135, 181, 203, 219, 228, 256, 21, 34, 77, 319, 374, 382, 675, 684, 717, 864, 203, 4, 18, 92, 16, 63, 82, 22, 46, 55, 69, 74, 112, 135, 186, 175, 119, 213, 416, 312, 343, 264, 119, 186, 218, 343, 417, 845, 951, 124, 209, 49, 617, 856, 924, 936, 72, 19, 29, 11, 35, 42, 40, 66, 85, 94, 112, 65, 82, 115, 119, 236, 244, 186, 172, 112, 85, 6, 56, 38, 44, 85, 72, 32, 47, 73, 96, 124, 217, 314, 319, 221, 644, 817, 821, 934, 922, 416, 975, 10, 22, 18, 46, 137, 181, 101, 39, 86, 103, 116, 138, 164, 212, 218, 296, 815, 380, 412, 460, 495, 675, 820, 952.

How did Ward manage to decipher the second paper? Since the number of numbers in the ciphertext greatly exceeds twenty-six (the number of letters in the alphabet), he wondered whether the numbers might correspond to the words in a document that Beale had sequentially numbered. With this in mind, Ward tried numbering the letters of words in many famous documents and substituting those letters for the numbers in the ciphertext. "All to no purpose," Ward wrote, "until the Declaration of Independence afforded the clue to one of the papers and revived my hopes." What Ward did was number the first letter of each word in the Declaration of Independence. For example, he numbered the first nine words as follows:

1   2   3   4   5   6   7   8
WHEN, IN THE COURSE OF HUMAN EVENTS, IT
9
BECOMES

From those words, he found that $1 = W$, $2 = I$, $3 = T$, $4 = C$, $5 = O$, $6 = H$, $7 = E$, $8 = I$, and $9 = B$. Already you can see that Beale had two ways of enciphering the letter I, as 2 or 8. Of course, by the time he numbered the whole Declaration of Independence, he had many choices for many of the letters. By making liberal use of all these choices, he made the ciphertext resistant to code cracking by frequency analysis. Ward was able to break the cipher because he stumbled on the appropriate keytext, the Declaration of Independence, with which he extracted the following message:

"I have deposited in the County of Bedford about four miles from Buford's in an excavation or vault six feet below the surface of the ground the following articles belonging jointly to the parties whose names are given in number three herewith. The first deposit consisted of ten hundred and fourteen pounds of gold and thirty eight hundred and twelve pounds of silver deposited November eighteen hundred and nineteen. The second was made December eighteen hundred and twenty one and consisted of nineteen hundred and seven pounds of gold and twelve hundred and eighty eight pounds of silver, also jewels obtained in St. Louis in exchange to save transportation and valued at thirteen thousand dollars. The above is securely packed in iron pots with iron covers. The vault is roughly lined with stone and the vessels rest on solid stone and are covered with others. Paper number one describes the exact locality of the vault so that no difficulty will be had in finding it."

Intrigued by this message, particularly by the last line, Ward devoted more and more energy to the decryption of the other papers. But try as he did, he didn't make any more progress. "In consequence of the time lost," Ward wrote, "I have been reduced from comparative affluence to absolute penury, entailing suffering upon those it was my duty to protect; and this, too, in spite of their remonstrances. My eyes were at last opened to their condition, and I resolved to sever at once, and forever, all connection with the affair, and retrieve, if possible, my errors. To do this, and as the best means of placing temptation beyond my reach, I determined to make public the whole matter, and shift from my shoulders my responsibility to Mr. Morriss."

And so, in 1894, he published an account of the Beale ciphers, an account that serves today as our sole source of knowledge of the ciphers and the fabulous treasure to which they supposedly lead. Every titillating detail that I have related to you—that Beale was tall and swarthy, that Morriss hit it off with both the rich and the poor, that savage animals accosted Beale and his hunting party—comes from Ward. There is no independent confirmation: no corroborating correspondence, no journals, no wills, no references to the treasure whatsoever. Moreover, the box Beale supposedly gave to Morriss has not survived, nor have the letters and enciphered papers that were allegedly in it. If Ward was a prankster, he was an extremely good one, having pulled off one of the longest-running hoaxes as well as one of the most expensive, if you consider all the computer time spent on unraveling the ciphers. "With computers," says Hammer, "we have played games with these numbers that would take a million men a billion years to duplicate with paper and pencil."

In the 1960s, some of the best minds in cryptoanalysis (and many of the worst ones too) formed a secret society, the Beale Cypher Association, so that they could pool their knowledge and resources and unearth the elusive treasure. Hammer, a prominent member of the association, ran extensive statistical tests on the distribution of the numbers in the undeciphered Beale papers and concluded that the numbers are not random but definitely conceal an English message. Most cryptographers accept Hammer's analysis, but the mere fact that there is a message doesn't mean that whole thing isn't a hoax. Who's to say that the message isn't something like "You're the biggest sucker in the universe, dodo brain"? Louis Kruh, president of the New York Cipher Society, performed some statistical tests of another kind, aimed at comparing the style of Ward's writing with the style of Beale's letters as they were quoted in Ward's pamphlet. Kruh found that the two styles are so remarkably similar that he is convinced that Ward wrote Beale's letters. For example, the average length of one of Ward's sentences is 28.82 words; that of Beale's, 28.75. Kruh's analysis, however, has caused few members of the Beale Cypher Association to turn off their computers and put away their shovels.

In 1981, the Beale legacy was given new life by Warren Holland, Jr., a dreamy, disenchanted graduate of Virginia Tech. Holland had not been able to make it in the construction business, because he had had trouble collecting money from clients. "In that business," says Holland, "forget about being honest, forget about being an individual. People push you into things you're not." With his mood and bank account depressed, he turned inward and read more than ever, including an account of the Beale ciphers and the scores of treasure hunters who, some 160 years later, were still digging up the backwoods of Virginia. Though intrigued by the story, he was not the sort who was about to race out and dig up the countryside—he'd done enough of that in the construction business. And then, eureka, it came to him, a way he could satisfy his own pecuniary needs by romantically appealing to those needs in others. He would encrypt his own message, market it, and offer a prize for its decryption.

It took him only a few hours to encrypt one of his favorite poems by e e cummings, called "A Poet's Advice," which spoke of the virtues of being nobody but yourself in a world that's doing its best to make you like everyone else. Holland proceeded just like Beale. First, he chose a key, not the Declaration of Independence but the sixth chapter of Carl Sagan's *Cosmos.* Then he numbered words consecutively, starting with the word *first* in a quotation at the beginning of the chapter, each number standing for the initial letter of a word. Finally, he substituted the numbers for the letters in "A Poet's Advice." He decided to write the encrypted text on a jigsaw puzzle, so that he would have a puzzle within a puzzle.

That was the easy part, an afternoon's work. The hard part, marketing the puzzle, took two years. He wanted to offer a prize of $100,000, which he planned to raise from the sale of the puzzle. But he wanted to insure the prize, in case the proceeds didn't amount to $100,000. Lloyd's of London turned him down because Scotland Yard claimed that the cipher could be easily broken. Eventually, he convinced an American insurance carrier and found a distributor for the puzzle. Called

Decipher, it sold nearly a quarter of a million sets in the two years it was marketed, until it was solved in March 1985.

In the winter of 1984, Alan Sherman, a twenty-seven-year-old Ph.D. candidate in computer science at MIT, decided to teach a minicourse in cryptography in which the goal was to solve Holland's Decipher puzzle. Six students took the course, including Robert Baldwin, a fellow graduate student. The class was armed with a Symbolics 3600 Lisp Machine, one of the most sophisticated personal computers available, and all the other resources of MIT's Laboratory for Computer Science, where Sherman had his office. (He now works four subway stops away at Tufts University, where he is an assistant professor.)

Uncracked ciphers were not strangers to the Laboratory for Computer Science. Many of the professors there have made monumental contributions to cryptography, although their concerns are generally more academic and theoretical than the pursuit of prize money for solving a commercial puzzle. And yet papers posted on the laboratory's walls suggest that the place isn't devoid of frivolity. Posted prominently is the July 10, 1984, front page of the supermarket tabloid *Weekly World News*, which features the story "Jealous Computer Kills Top Scientist: Old Machine Electrocutes Owner—After He Buys a More Advanced Model." Also tacked to the wall are curious block-by-block maps of various Soviet cities. The CIA used to have offices on the same floor, and when the agency moved out of the building, MIT students fished the maps out of a dumpster along with a pamphlet called something like *How to Trail People in the City.*

Sherman himself almost succumbed to the allure of skulduggery, not the cloak-and-dagger espionage practiced by the CIA but the genteel computer-keyboard snooping practiced by the

NSA, the chief code-breaking and code-making wing of the government. Even the budget of this stealthiest of governmental organizations is classified, although it's thought to be twice that of the CIA. Its operations are so hush-hush its employees joke that NSA stands not for National Security Agency but for Never Say Anything. The NSA was responsible for the controversial Data Encryption Standard, a complicated cipher that other government agencies and private companies were supposed to use in order to keep information confidential in files they maintain on private citizens. Critics charge that NSA advocated the cipher, touting it as virtually impenetrable, because the agency built into it a secret trapdoor that it could effortlessly activate whenever it wanted access to confidential records. Sherman is not one of these critics, but he has devoted much energy to investigating the mathematical properties of the Data Encryption Standard and how those properties relate to the security of the cipher. He has probed one curious property of the cipher: there are messages that are identical to their encryption!

When people leave academe for the NSA, it's generally not because they're overcome by a sudden urge to serve their country. Rather, it's because the agency appeals to the techno-nerd in them: the NSA's top-secret facility in Maryland apparently houses more computers than are found anywhere else on the planet. Sherman turned down the agency's job offer because its strict security regulations might have kept him from ever teaching cryptography again or from publishing papers on the subject. Given the agency's notorious gag orders, it is conceivable that an NSA employee has solved the Beale ciphers but is barred by agency rules from reporting his solution or digging up the treasure, even under cover of darkness.

When I met Baldwin, in the spring of 1985, he was interested in "cryptographic protocol," the use of cryptography to achieve "higher goals." I should have asked him what the lower goals were; all I could think of was the NSA's reported interception of a radio call between Moscow limousines, in which a Kremlin bigwig revealed the special services of a local masseuse. Baldwin, though, needed no prodding to discourse at length on the higher goals. He told me how cryptographic signatures could be used so that when you communicate from a keyboard with a computer, you know it's the computer you're interacting with—and not some nefarious high-tech saboteur who's impersonating the machine. Another higher goal is the encryption of personal checks and credit-card receipts so that no one but the customer knows what he's spent his money on. "Checks should be anonymous," says Baldwin. "They should not provide a trail of where you've been. It's nobody's business if you write a check to your mistress."

Baldwin treated me to a demonstration of the computer system that they used to attack Decipher. He started the program, and the screen filled up with the following text:

WARNING: PERMISSION TO USE THIS SYSTEM IS ONLY GRANTED TO ITS PRIMARY IMPLEMENT-ERS. ARE YOU AN IMPLEMENTER?

"We didn't do anything sophisticated to keep out unauthorized people," says Baldwin. "Just this, which makes the moral choice very clear."

The idea of the system is that the user types in a candidate keytext and the computer tests all sorts of strategies for assigning numbers to that text. One strategy is to number the first letter of every word, as Beale did with the Declaration of Independence. Another way is to number every letter. And each of these strategies is tried again and again, starting at various points in the text. Each way yields a different assignment of letters to numbers, which the computer then applies to the ciphertext in an attempt to extract an English message.

Since the computer doesn't read English, Baldwin and Sherman had to build in a method it could use to tell whether the extracted text was gibberish or possibly an intelligible message. They did this by having it conduct a statistical test. It counted the frequency of certain letter pairs in the extracted text and compared this with known frequencies for English. If the frequencies were close, the computer would store the extracted text for the perusal of its more literate human masters.

So far, so good. But the success of the system depended on Baldwin and Sherman's typing in the right keytext. In this respect, even with all the resources of modern cryptography, they were no better off than Ward. Indeed, Ward may have had it easier because there were fewer documents in print in 1820 and thus fewer candidate keytexts to consider. Holland, however, gave out a few cryptic clues: "3, 19" and "If you knew it began with C, would it help you?" The first clue was supposed to reveal Carl Sagan's initials since C is the third letter of the alphabet and S the nineteenth. The second clue applied to *Cosmos,* since it begins with C. As the months passed and no one solved Decipher, Holland gave out increasingly helpful clues over a telephone hotline that puzzlers were urged to call.

"In the beginning of March 1985," Baldwin recalls, "Holland released a clue that the key was a sequence of first letters from chapter 6 of *Cosmos.* We deduced that he meant first letters of words because there weren't enough lines or sentences for it to be first letters of those. We hired somebody to type in the chapter, and by the middle of March we were running our program but we weren't getting a match. We actually tried the strategy, which Holland actually used, of beginning with the word *first,* numbering it 1, and so on. We were right up to the 256th word, which was *c.*, the abbreviation for *circa.* We figured that Holland would count the *c.* as a word since it stood for a word. In fact, he deleted it. That meant that everything else we numbered was off by one—and that small difference produced total gibberish.

"There were other peculiarities. At one point Sagan writes JPL, for Jet Propulsion Laboratory. Does that count as one word or three words? Holland chose to delete it. The c. and JPL as well as other complications—acronyms, footnotes, picture captions, hyphenated words, and numbers in the text—defeated our program. When we began, we were thinking in terms of a keytext like the Declaration of Independence, which has few of the complications of a modern document like *Cosmos.* It wasn't until late in the game that we thought well, gee, our program is pretty careful about trying thousands of different strategies—taking every letter after a vowel or any other weird thing we could think of—but it wasn't very clever about massaging the text, of deciding what's a word and what isn't. We realized that there were about sixty different ways to treat the acronyms, footnotes, hyphenated words, and other complications. The program wasn't designed to do that, and we weren't about to try them all by hand."

On March 27, Sherman and Baldwin developed a clever method for detecting partial matches between part of the message and part of the keytext—a method that avoids the problem of how to treat the peculiarities of the text. They noticed that at various places in the ciphertext adjacent cipher symbols were numerically close. For example, at one point there was the sequence 867, 877, 860—numbers that differ by at most 17. By concentrating on instances of seventeen consecutive words in the keytext that were free of textual complications and numbering the words consecutively from 860 to 877, they could extract plaintext letters for 867, 877, and 860. They actually did this on a more elaborate scale so that they could extract enough plaintext letters to be able to analyze them statistically. As before, the program compared letter-pair frequencies in the extracted text with known statistics for English as a whole.

On March 29, the computer found an extraction that exhibited the proper statistics. Bob went to work filling out the partial match until he had the complete text of e e cummings's verse. "It's funny," he says, "but 99.8 percent of all English texts would score better statistically than the poem in terms of being closer to average English. For example, cummings used the words *no* and *you* fifteen times. Yet the poem is still so much closer to English than to non-English. We were lucky that we had enough slack in the statistics." If they thought e e cummings was bad, they're lucky Holland's favorite poet wasn't Gertrude Stein. What would their program have made of the letter frequencies in "Rose is a rose is a rose"?

Unfortunately for Sherman and Baldwin, the deadline for submitting a solution to the Decipher contest wasn't the last day of March, as they had thought, but the last *business* day. "It's incredible we didn't realize that," says Baldwin. "We figured we're MIT students, so we don't have to carefully read the rules." They take some consolation in the fact that they wouldn't have won the $100,000 outright but would have had to split it with thirty-six other puzzlers who had submitted timely solutions. "Besides," says Baldwin—never one to pass up a chance to calculate—"we had promised to give half of our share to the university for use as financial aid. And the other half would be split among Alan, the typist, and me. The typist, you see, was working for a percentage because we couldn't afford to pay him. That works out to my share being $700. Hell, I can earn that much money doing consulting work for two days." And if the consulting business dries up, Baldwin can always go after Beale's treasure trove by training their computerized code-cracking system on his ciphers. Of course, there is the nagging problem, so far immune to advances in cryptography and computer number crunching, of identifying Beale's keytexts—and, unlike Holland, Beale is not around to give out clues.

II

SHAPES

A rudimentary knowledge of mathematics certainly antedates the invention of the wheel, the use of metal, and the development of writing. Prehistoric artifacts indicate that early man counted, with the aid of notches in a tally stick. For example, a wolf bone unearthed in Czechoslovakia contains fifty-five deep notches, made in groups of five some thirty thousand years ago. The ancient cultures of Egypt and Mesopotamia knew about geometry as well as arithmetic, although there are only fragmented records of exactly what they knew. Aristotle believed that it was Egyptian priests, with leisure time to pursue intellectual endeavors, who developed geometry. But Herodotus, a Greek historian, thought that geometry was developed in Egypt out of necessity. Each year the Nile seriously flooded the farmlands in the river basin, washing away property markers. To resurvey the land, as the Egyptians did annually, required an understanding of angles, directions, and lengths.

Geometric knowledge was also undoubtedly required to construct the great Egyptian pyramids.

Most ancient and primitive cultures seem to have an appreciation of geometric form. Indeed, a well-developed sense of shape may be part of the human psyche. Even "the man who mistook his wife for a hat"—Oliver Sacks's neurologically impaired musician who couldn't recognize faces and ordinary objects—could recognize geometric forms. As a mathematical discipline, geometry is as alive today as it ever was, and geometers are still making discoveries about the simplest of shapes.

5

ADVENTURES OF AN EGG MAN

"It's the best idea I've ever had," says Kay McKenzie, an alderman in Vegreville, Alberta. She was speaking of her plan to commission a three-and-a-half-story Easter egg for the barren, tornado-swept field opposite the nursing home in this sleepy agrarian town fifty-five miles east of Edmonton. Although McKenzie herself is not Ukrainian, most of the five thousand people in Vegreville are, and they still practice the two-millennium-old Easter tradition of painting *pysanki*, brightly colored, intricately patterned chicken eggs. To celebrate the centennial of the Royal Canadian Mounted Police in 1974, the Canadian government was offering grants for appropriate community projects. Why not a giant egg? McKenzie thought. It would symbolize the peace and security that the Mounties had offered to generations of Ukrainians in Vegreville.

At first, her colleagues in town government roared with laughter, but ultimately she was able to persuade them that the egg was an idea whose time had come. After all, they imagined the grant committee welcoming a fresh idea after reviewing countless proposals for statues of Mounties on horseback, trumpeting Canada geese, and golden maple leaves. In fact, many of the submitted proposals were hopelessly banal. There were numerous plans for refurbishing old buildings and, almost as an afterthought, slapping a plaque on the wall in tribute to the Mounties. In the end, Vegreville received $15,000—and a matching grant from the local chamber of commerce—and immediately sought an egg builder.

The town leaders asked a respected local architect to build them the world's largest decorated chicken egg—and he roared with laughter. After a few months, they called him up to check on his progress; he reported that he hadn't done any work, because he thought they were putting him on. They tried another architect, who laughed even louder. Six design firms later, they contacted Ronald Dale Resch, then a thirty-five-year-old associate professor in computer science at the University of Utah. "At first," recalls Resch, "I laughed too, but when they ended up giving me the job, I quit laughing for a year and a half."

The problem Resch faced was that no one other than a chicken had ever built an egg—and biologists are not all that clear about how chickens do it, some 390 billion times a year, according to the trusty *Britannica*. In the domestic hen, an egg takes about twenty-four hours to form, beginning as yolk (the ovum) in the hen's ovary. The incipient egg starts a long, stop-and-go trek through the oviduct, the tubular passage that leads from the ovary to the vagina. On the egg's first, three-hour rest stop, it picks up albumen (egg white) secreted from cells in the oviduct walls. The egg then inches along to a section of the oviduct where, pausing for an hour, it receives the membranes that will line the shell. Finally, it moves on to the uterus, where over a period of twenty hours, it accumulates chalky deposits that harden into the shell. So far, the egg has always traveled with its more pointed end leading the way. But half an hour before emerging, the egg flips over so that it's laid blunt end first.

The egg is initially a fluid structure. In the absence of external forces, the egg would be a sphere, a shape that minimizes contact with the rest of the world. Given a certain volume of fluid, of all shapes that could contain that volume, the sphere has the smallest surface area. The eagle owl and the kingfisher actually lay eggs that are nearly spheres, but most birds are like the chicken; their eggs are elongated because the muscular contractions of the oviduct, which propel the eggs by squeezing them, modify their spherical form.

Virtually all forms in nature serve a function, and the shape of an egg is undoubtedly no exception, even if science cannot yet identify the shape's precise function. Perhaps it has something to do with how the egg rolls. If a chicken egg were a sphere, it would be much more prone to rolling away. Certain sea birds, such as the guillemot, a narrow-billed auk that inhabits northern waters, lay eggs that are even less like spheres than chicken eggs are. The guillemot egg is shaped like a top, whose dynamics are such that, when it rolls, it moves not in a straight line but in a tight circle. This is fortunate for the guillemot, who is more of a daredevil than a home builder; spurning a nest, the guillemot lays its top-shaped eggs directly on the slippery edges of ocean cliffs.

That the eggs of chickens and many other birds are broader at one end than the other means that they can be packed together closely in a nest—more so than if they were spheres. "If the four eggs in the nest of a killdeer [a North American plover known for its woeful, penetrating cry] are disarranged," writes Joel Carl Welty, an ornithologist at Beloit College, "the bird will rearrange them with pointed ends inward much like the slices of a pie. Not only is the parent better able to cover its eggs, but the heat they receive from its body is dissipated less rapidly, thanks to their compact positioning."

Perhaps the shape of the egg also contributes to its strength. After all, it must not break under the weight of a nesting parent. Given its size and the thinness of its shell, a chicken egg may be relatively strong, but not so strong that it will survive, as legend has it, the grip of a muscleman who squeezes it longitudinally in one hand. Perhaps this is the same mythical muscleman who can rip a phone book in two. (The egg's legendary strength has been touted by a recent advertisement that pictures an unbroken egg gripped by a metal C clamp.) In reality, you don't even have to be a he-man to crush an egg single-handedly; I did it with one of my grubby little mitts at the age of six, thereby advancing science but setting back the kitchen floor.

"It may be true in theory," says Resch, "that if a strong man applied pressure *uniformly* to the surface of an egg, he might not be able to break it. In practice, however, no one applies uniform pressure—it's greater at some point than another—and the egg bursts. In textbooks, they like to show a whole bunch of eggs, with plaster above and below them, and elephants standing on them without crushing them. That only shows what's true of any structure: if you distribute a load properly, it will carry it. In the real world, however, loads are never distributed properly."

That Resch thinks about what works not only in theory but also in practice made him the ideal egg man, one who could see the giant egg through its incubation from a design on paper to an awesome monument reaching up thirty-one feet and weighing two and a half tons. Resch lives by the simple motto "Have mind, will travel." Sometimes he leaves the United

States for months at a time to meditate in India. At other times, he sets up shop near a university or research center and works on his geometric art and computer graphics. But mostly he moves around, hiring himself out to people, like the folks in Vegreville, who need help solving a tough problem in geometric design. With no formal training in mathematics or engineering, Resch relies chiefly not on analytic methods but on his ability to play with geometric abstractions in his mind's eye and then, with his own hands (or, these days, with his computer printer), to turn those mental abstractions into physical reality.

For NASA's Langley Research Center, in Virginia, he designed prefabricated modules that, fitting snuggly in the space shuttle's cargo bay, could be carried into space, deployed, and then linked together to form huge space-station structures. The producers of the film version of *Star Trek* hired him to design the mouth of an alien vehicle; they told him to make it look both organic and high-tech, and he came up with the techno-mouth of the mysterious spaceship that swallows everything in its path, including the starship *Enterprise*. For Royal Packaging Industries Van Leer, a Holland-based multinational package-design conglomerate, he devised a more efficient way of packing in crates such spherical fruits as apples and plums.

Finding the densest way to pack various geometric objects is an age-old problem in mathematics that has engendered much discussion. In 1694, for example, Isaac Newton argued with the Oxford astronomer David Gregory about the maximum number of spheres, all identical in size, that could be in contact with any one sphere of the same size. Gregory said thirteen and Newton twelve. One hundred eighty years passed before Newton was proved right.

Kissing Spheres

The packing of twelve spheres around a thirteenth sphere is the key to the packing of spheres in the densest known way. Imagine a bunch of spheres in a straight line on a flat surface like a desktop. Now put another line of spheres up against the first line so that the spheres of one line fall between the spheres of the other; any given sphere should be kissing two spheres in the other line. More lines of spheres should be added until the desk is covered. A second layer of spheres should be created by adding them to spaces between spheres in the first layer. A third layer is then formed by positioning spheres in spaces in the second layer. If this kind of layering is not restricted to the desktop but fills all of space, the spheres will occupy 74 percent of the space. In other words, it's necessary to waste 26 percent of the space. No one knows whether a denser packing exists.

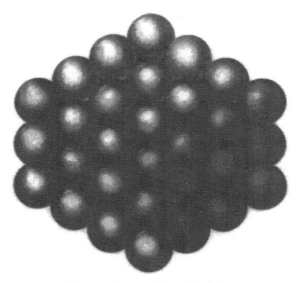

Densest Known Sphere Packing

When Resch started the apples-and-plums consulting job for Royal Packaging Industries Van Leer, he assumed that when spherical fruits are shipped in rectangular crates, they should be packed according to either of these two densest known arrangements. For months he proceeded that way, until it suddenly occurred to him that the densest known packings made the mathematical assumption that the entire cosmos was filled with spheres. But here, in the real world, he was dealing with a small finite volume, a three-foot-by-four-foot crate. With that insight, he was on the road to a solution, but he had learned an important lesson: the world itself imposes all sorts of constraints that armchair theorizing never addresses. (Resch refuses to reveal his solution because it's not patented.)

Resch is fond of saying that design is "a kind of feedback loop between the designer and the environment"—a description that also applies to his own career. Raised in Independence, Missouri, Resch thought about going into professional sports. In high school, he was a three-letter athlete, in football, basketball, and track, but a heart murmur, detected in his junior year, forced him to abandon sports altogether. "I was always good with my hands," says Resch, who channeled all the energy he could no longer expend on the playing field into art, particularly sculpture, for which he received a scholarship to the University of Iowa.

Once at Iowa, he also studied industrial design and stayed there until he got an M.F.A. degree in 1966. But with no technical training in engineering, he couldn't get a job in industry. "All these corporate types disapproved of the fact that I hadn't taken any math courses," Resch recalls. "At the time I couldn't avoid an onslaught of cultural judgment that I was nothing. Today, though, I feel vindicated. I can build things, unlike the brilliant idiots the schools are churning out as engineers, who know all the abstractions but can't do the nuts and bolts. It also pleases me that today geometric design is critical to so many major efforts in physics, chemistry, and computer science."

Resch's approach to design is to take certain basic, minimal forms and examine all the ways they can be manipulated into more complicated structures. "I have made a profession," says Resch, "of studying one of the simplest forms, the single sheet," and seeing what happens when it's bent and folded every which way. "It's not origami," says Resch, "where the intent is to produce a recognizable shape. I'm interested in creating systematic modular forms." And that is what he has done—some would say to the point of excess. For over two decades, he has been manipulating single sheets (of paper, aluminum, and other materials) into three-dimensional forms that exhibit some kind of pattern or regular structure. He has shown the more interesting ones in art galleries and, somewhere along the way, acquired a patent on what he believes, but can't prove, are all the possible ways of folding a sheet into a repeating pattern.

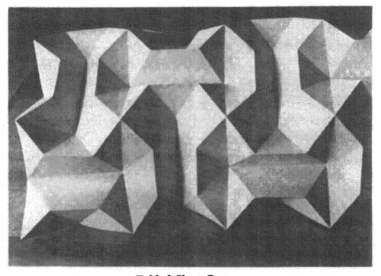

Folded Sheet Pattern

"I took on the egg project," says Resch, "because I thought it would be easy. At the time, I had just made one of my folded-sheet structures in the shape of a dome. It looked much like the end of an egg, so I thought I'd take two of these domes, put a kind of bulging barrel shape between them, tie the three together," and, presto, an egg. Resch had developed a computer program to do simulations of folded-sheet structures, and he thought that, with only a slight modification, it could simulate an egg. "When I took on the job," Resch recalls, "I assumed that surely someone in the history of mankind had developed the mathematics of an ideal chicken egg." By comparing the mathematics with the geometry of his simulation, he expected to be able to judge analytically how good the simulation was.

Resch soon found, however, that there was no formula in the literature for an ideal chicken egg. For many shapes that have a name, the literature contains not only an algebraic formula but also a method of construction. Take the circle. It is simply the set of all points in a plane that are equidistant from a given point in that plane. To construct a circle, tie one end of a length of string around a pencil and anchor the other end with a thumbtack to a piece of paper. With the string pulled taut and the pencil point held against the paper, rotate the pencil around the thumbtack; the result is a circle. At some point, a twisted wag even turned this simple construction process into a sick joke, which I learned from the mathematician Martin Gardner: "Mommy, mommy, why do I always go 'round in circles?" "Shut up, kid, or I'll nail your other foot to the floor."

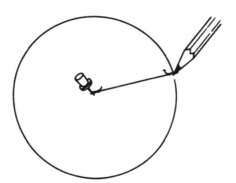

Construction of a Circle

It's an easy step from a circle to a sphere—imagine the kid's foot (or the string's end) nailed to a point in three-dimensional space, swing the kid's rigid body (or the pencil on the end of the taut string) every which way and observe the shape the kid's head (or the pencil point) traces out. Alternatively, you can think of the sphere as the shape swept out by a pirouetting circle.

A chicken egg, of course, is closer to an ellipsoid—the shape swept out by a pirouetting ellipse—than to a sphere. Even the most demented mathematician wouldn't be able to generate an ellipse by twirling a child but could do so easily with the aid of a pencil and a loose string anchored by thumbtacks at both ends.

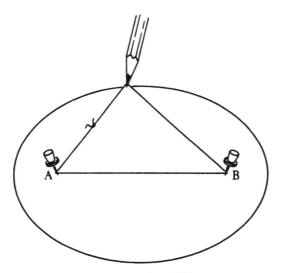

Construction of an Ellipse

Unlike an ellipse, a chicken egg is blunter at one end than at the other, but this asymmetry doesn't mean it can't be represented mathematically. Indeed, back in the seventeenth century, the French man of letters René Descartes ("I think;

therefore, I am") explored the algebraic formulas for egg-shaped curves. Two centuries later, the Scottish mathematical physicist James Clerk Maxwell, best known for his quantitative demonstration that electricity and magnetism are part of the same phenomenon, extended Descartes's efforts. Maxwell was merely fifteen at the time, and he sent off a paper on egg shapes to the Royal Society of Edinburgh, Scotland's premier scientific society. The paper was warmly received, but the august society refused to let such a pip-squeak address them on the subject. It missed an arresting demonstration that egg-shaped curves can be constructed with pencil, thread, thumbtacks, and a little ingenuity.

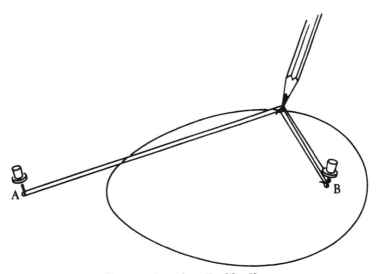

Construction of an Egglike Shape

The string is anchored initially at point B, then wrapped twice around the pencil and once around the thumb tack at point A. The end is then tied to the pencil. With the string taut, the top half of the egg shape can be traced. The string and pencil setup is then inverted so that the bottom half can be drawn.

Resch's main problem was that if you've seen one chicken egg, you haven't seen them all. They do vary slightly in shape, and it was up to him to discern the ideal form. In a fit of frustration, he called up the agriculture department and had it airfreight him an egg-grading manual. "I thought," says Resch, "that the manual would surely include a definition of a chicken egg. But all I found were photographs labeled A, AA, B, and BB. Finally, I came to an image called the ideal egg. So I had it photographed and then digitized in my computer program." For six months, Resch and two graduate students worked day and night to turn his folded-sheet structures into an egg, but all they got were negative results. "We didn't know what was wrong—our program, our geometry, or our mathematics."

Resch ended up throwing out his computer program, setting aside the folded sheets that had served him so well for two decades, and starting over from scratch. His plan was to construct the egg out of numerous flat tiles joined together at slight angles, treating the egg as if it were a three-dimensional jigsaw puzzle. In theory, infinitely many different configurations of puzzle-piece tiles would do the trick, but Resch needed more than a mathematical solution. His tiles would have to be machined, and so, in the interests of economy, it was important that as many of the tiles as possible be of the same shape and size; that way they could be cut from the same mold.

In two dimensions, tessellations—tile patterns in which a flat surface is completely covered by tiles (straight-line shapes) that do not overlap—have a long and rich history. The third-century astronomer Pappus of Alexandria, in marveling at the geometric structure of a honeycomb, attributed to bees "a certain geometrical forethought" in building hexagonal (six-sided) cells. Since hexagon tessellate the plane, beeswax is saved because each wall is always common to two cells. More-

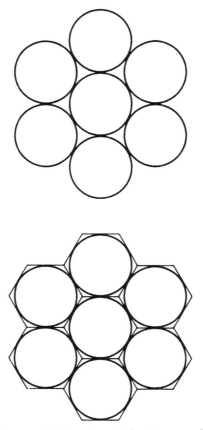

Hexagonal Advantage of the Honeycomb

Of all two-dimensional shapes, the circle has the maximum area for a given perimeter, but it is not suitable for a bee cell, because too much space would be wasted between circles. Another advantage of the hexagonal form derives from the sharing of adjacent sides. The six outer hexagons yield the inner one for free, since sides are shared, but the six outer circles do not yield a free inner circle—it must be drawn— since circles never have sides in common. More subtle is the savings that comes from the sharing of adjacent sides by the six outer hexagons. The six are constructed from the perimeters of only five hexagons. Although seven circles are indeed seven circles, five hexagons are effectively seven.

over, Pappus thought it splendid that "no foreign matter could enter the interstices between [honeycomb cells] and so defile the purity of [the bees'] produce." Pappus observed that besides the hexagon, the square and the equilateral triangle are the only other regular polygons (straight-line figures with all sides equal and all angles equal) that can tile the plane by meeting corner to corner, but for the bee the hexagon is superior because it encloses the most area for a given perimeter. In other words, of the three shapes, it holds the maximum amount of honey for the minimum expenditure of wax.

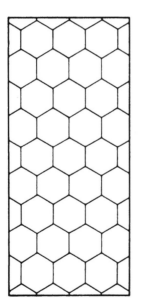 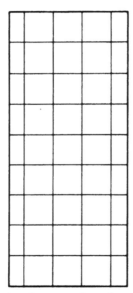 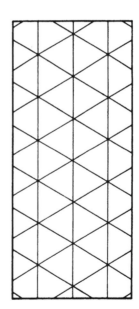

The Three Regular Tilings

It is easy to convince ourselves that Pappus did not over-look any regular polygons that tile the plane. The key condition is that the polygons be able to fill up space about a vertex. To do this requires six triangles, four squares, or three hexagons. These three kinds of polygons can surround a vertex because the interior angle (60 degrees for the triangle, 90 degrees for the square, and 120 degrees for the hexagon) divide evenly into 360 degrees. No other regular polygon has this property. A regular pentagon, for example, has an interior angle of 108 degrees, and so when three pentagons are placed around a vertex, 36 degrees of the floor remain untiled.

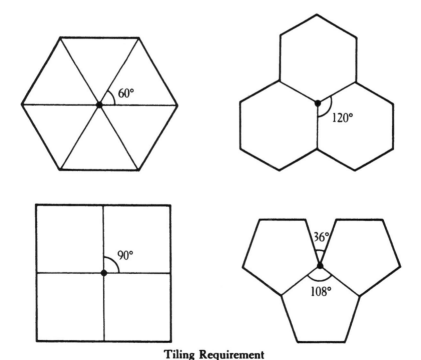

Tiling Requirement

If the requirements are relaxed so that more than one kind of regular polygon is allowed in a tiling, but all vertices are identical (in the sense that the order in which polygons are placed around any one vertex is the same as that around any other vertex), eight additional tilings are possible. Depending on whether you're mathematically or empirically inclined, try to convince yourself, either by armchair analysis or by a comprehensive survey of fancy bathroom floors, that no other such tilings are possible.

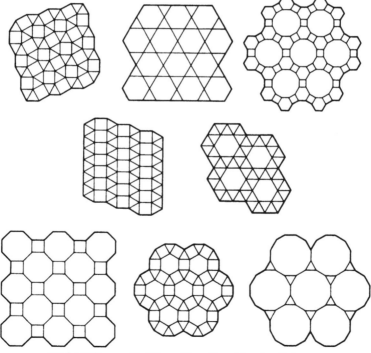

Eight Tilings with More Than One Regular Polygon

The tilings we have mentioned so far are all periodic; they repeat like wallpaper. Each tiling contains a "seed," the smallest unit that the tiling as a whole is a multiple copy of. If you had a rubber stamp of the seed, you could create the entire tiling by repeatedly using the stamp, moving it up or down or side to side, never needing to rotate it. In the three tilings consisting of only one regular polygon (the triangle, the square, and the hexagon), the seed is obviously the polygon itself; the honeycomb tiling is created from a single hexagon, the square tiling from a single square, and the triangular tiling from a single equilateral triangle. The Dutch artist M. C. Escher is well known for his periodic tilings, in which the tiles are generally not regular polygons but animals of one sort or another.

For a tiling to be nonperiodic, it need not be complicated. Picture a square tiling. Now imagine each square split along one of its diagonals into two right triangles. It's up to you which diagonal to split each square along, but all the squares must be split in such a way that the overall tiling of right triangles is nonperiodic. This nonperiodic tiling could not be simpler: it consists of only one kind of tile—a right triangle—and even though the tiling does not have a seed, it is predictable in the limited sense that the triangles form squares.

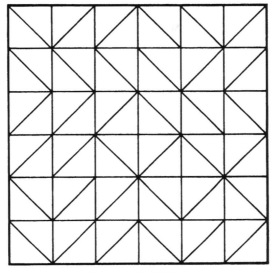

Nonperiodic Tiling

With a minimum of effort, the right triangles in this non-periodic tiling can be rearranged so that the tiling is periodic. An easy way to do this is to rotate 90 degrees each two-tile square in which the diagonal runs from the upper left to the lower right. That way all the diagonals will run in the same direction, and the seed is simply the two triangular tiles that make up any square.

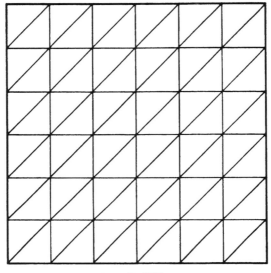

Periodic Tiling

A nonperiodic tiling can be created from any number of different kinds of tiles. That the number of possibilities is unlimited makes nonperiodic tiling the choice of geometrically minded snobs who want their bathroom floor to be unique. To create a nonperiodic tiling from two kinds of tiles, we could also start with squares but, instead of splitting them diagonally, we take a right-triangular nick out of either the northwest corner or the southeast corner. As in the earlier case, there should be no pattern to which of the two corners we choose, and all the nicks should be the same size. The result is a nonperiodic tiling made up of right triangles and irregular pentagons. Again, the

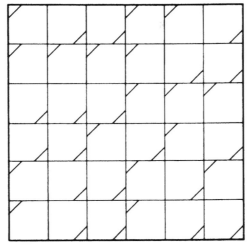

Nonperiodic Tiling of Two Tiles

tiles can be rearranged into a periodic pattern by, say, taking each square that has a triangular tile in the southeast corner and rotating it 180 degrees.

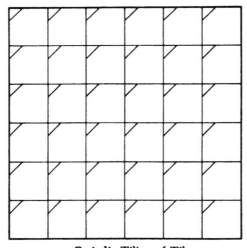

Periodic Tiling of Tiles

In the early 1960s, mathematicians believed, but couldn't prove, that for any nonperiodic tiling based on at least two different kinds of tiles, there was also a periodic tiling that involved the same kinds of tiles or a subset of them. In 1964, Robert Berger, a graduate student at Harvard, demonstrated that this belief was false. Ten years later, while Resch was at work on his egg, Roger Penrose, a theoretical physicist at Oxford whose imagination is unbounded, came up with two tiles—called kites and darts—that do the trick. The kites and darts must come together at corners, like the other tiles we

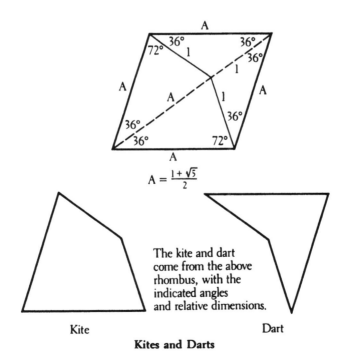

$$A = \frac{1 + \sqrt{5}}{2}$$

The kite and dart come from the above rhombus, with the indicated angles and relative dimensions.

Kite   Dart

Kites and Darts

have looked at, but certain sides are not allowed to be in contact with other sides. This restriction is enforced by putting bumps and dents on the tiles that keep them from being aligned in prohibited ways.

Bumps and Dents on Kites and Darts

Amazingly, the kites and darts can tile the plane in infinitely many ways, not one of which is periodic. The patterns may have a high degree of symmetry, but they always stop short of repeating themselves.

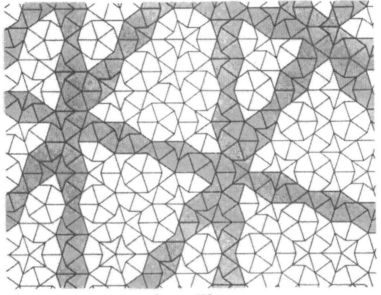

Penrose Tiling

(Reproduced courtesy of Longman Group, Ltd., U.K.)

What is most remarkable, any finite region in any one of these tilings comes up infinitely often elsewhere in that particular tiling and infinitely often in every other tiling. "To understand how crazy this situation is," writes Martin Gardner in a *Scientific American* cover story (January 1977)—must reading for devotees of Penrose tiles—"imagine you are living on an infinite plane tessellated by one tiling of the uncountable infinity of Penrose tilings. You can examine your pattern, piece by piece, in ever expanding areas. No matter how much of it you explore you can never determine which tiling you are on. It is of no help to travel far out and examine disconnected regions, because all the regions belong to one large finite region that is exactly duplicated many times on all patterns. Of course, this is trivially true of any periodic tessellation, but Penrose universes are not periodic. They differ from one another in infinitely many ways, and yet it is only at the unobtainable limit that one can be distinguished from the other."

And if this is not mind-blowing enough, Gardner goes on to explain another remarkable property, discovered by John Horton Conway, a mathematician at the University of Cambridge. Imagine you live in a town—a circular region of any size—somewhere in one of these Penrose universes. How far must you go to find an identical town? Conway proved that you never have to venture farther than twice the diameter of your town! Moreover, if you were suddenly transported to any of the infinitely many other Penrose universes, you would also always be at most two diameters away from a region that matches your hometown—and the odds are that you would be at most one diameter away.

The cosmological implications of Penrose's work are astounding. Out of only two simple primitive components—atoms, if you will—an unlimited number of universes can be built, all of which exhibit tremendous regularity on any conceivable finite scale but are uniquely irregular on the cosmic scale.

Resch's concerns were more down-to-earth, even though his project bordered on fantasy—an egg so large an army of Easter bunnies couldn't lift it. He knew that the extensive mathematical and architectural literature on tiling patterns applied only to flat surfaces, not to curved ones like that of an egg. Confronting the challenge of the unknown, he pictured an egg with latitude lines on it. In other words, he imagined the egg constructed from bands, stacked one on top of the other, each band individually tiled. But a computer simulation of this natural idea showed that even when the bands were thin and the tiles numerous, the human eye would fixate on the bands rather than taking in the shape as a whole.

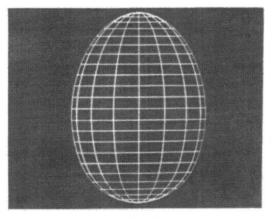

Banded Egg

Abandoning the bands, Resch turned to one of the simplest shapes, the equilateral triangle. After six months of contemplation and simulation, Resch realized that he could tile the egg with 2,208 equilateral triangles of identical size and 524 three-pointed stars (equilateral but nonregular hexagons) that varied slightly in width, depending on their position in the egg. The angle at which tiles were joined would vary from less than one degree in the bulging middle of the egg to a mere seven degrees at the pointed end. With such small angles, the egg would appear to curve smoothly, even though it consisted of flat tiles. Made of 2,000 pounds of anodized aluminum, the triangular tiles would be an eighth of an inch thick; the star-shaped tiles, half that. Held together by a 3,000-pound internal structure, the egg would be 25.7 feet long and 18.3 feet wide.

"Never before," says Resch, "had a three-dimensional surface like this been tiled with largely identical tiles. For example, the heating tiles on the space shuttle were all different. If the shuttle's designers had known about my work, or I had known about their problem, the shuttle might have been tiled like the egg. That way they could carry replacement tiles into space." As it was, the shuttle didn't carry spares, since each tile was unique. When the tiles fell off, as they often did when the shuttle sped through the atmosphere, new ones had to be machined.

"When Vegreville hired me," says Resch, "the deal was that I would design the egg and they would take care of building and painting it. It became clear to me, however, that there was no way Vegreville could build it without calling in an aerospace company to machine the tiles. And they certainly couldn't afford to do that. So I told them I'd build and paint it."

The painting of the tiles, which was done before they were assembled, involved some compromise. The town wanted the egg brilliantly decorated in vibrant reds, blues, greens, and oranges but expected the paint job to last a hundred years. With those colors, Resch told them, the egg would have to be repainted every three to five years. In the end, three colors were chosen—gold, silver, and bronze—that should retain their luster for half a century.

Before he could begin construction—joining the tiles internally, where the connections would be invisible, by 6,978 nuts and bolts and 177 struts connected to a central shaft—town regulations required a civil engineer or an architect to certify that the design was structurally sound. But mindful of the hundred-mile-an-hour winds that often swept Vegreville, none of the local engineers or architects would attest to the structural integrity of such a huge unfamiliar form. "There was fear," says Resch "that it might blow away. And I admit that

I, too, was a little scared. While building it, I'd be inside the damn thing and under it." At that point, the project had acquired a momentum of its own and the town simply waived the certification regulation. Many citizens of Vegreville bet among themselves not on *whether* the egg would collapse but on *how* (by toppling over or blowing away) and on when (during construction or afterward).

For six weeks, Resch led a team of volunteers in assembling the egg. They had one close call when the top part of the egg was completed and mounted on the end of the shaft so that it looked like a huge umbrella. A fierce storm kicked up, and a tornado dropped out of the sky. Resch and the others spent a long night turning their umbrella-like structure into the wind so that it wouldn't take off.

The egg had to withstand not only the forces of nature but also the wrath of man. After a hard day of egg building, Resch would collapse in one of the local taverns, where he'd overhear people muttering about plans to blow up the egg. He was even warned a few times that high school kids were going to dynamite it. He eventually discovered that moments before his arrival in Vegreville the newspaper had run a story erroneously claiming that town funds earmarked for building a swimming pool at the high school had been diverted to the egg. "I ran around," says Resch, "trying to explain to everybody where the money was actually coming from and that they'd get their swimming pool. No one ever tried to blow it up, but it did sustain a few rifle shots."

Long after the egg was finished, Resch used a computer to analyze its structural integrity and concluded that it was stronger than it needed to be by a factor of ten. "The whole community," says Resch, "would blow down before the egg would."

A decade has passed since Resch left Vegreville. The town is still standing and, consequently, so is the peculiar monument that put Vegreville on the map (and on Queen Elizabeth's Canadian itinerary). The town's only complaint is that the egg hasn't yet made it into the *Guinness Book of World Records*. It doesn't seem fair that Calgary, another town in Alberta, should be heralded in *Guinness* for such frivolity as cooking the world's largest omelet, from 20,117 eggs.

6

THE MÖBIUS MOLECULE

A mathematician confided
That a Möbius strip is one-sided,
And you'll get quite a laugh
If you cut one in half,
For it stays in one piece when divided.
　　　　　　　　　　—ANONYMOUS

Mathematics can aid in the design of shapes not only on the grandest scale, like that of a three-and-half-story Easter egg, but on the microscopic scale. This chapter is the story of how David Walba and his coworkers at the University of Colorado at Boulder synthesized a molecule in the peculiar shape of a Möbius strip.

The weird Möbius strip is the darling of mathematicians. You can make a Möbius strip by taking a narrow strip of paper, such as adding machine tape, giving it a half-twist, and joining the ends with tape to form a closed ring.

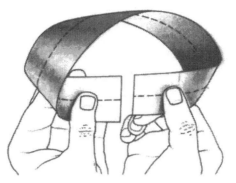

Making a Möbius Strip

Sure enough, if you cut the Mobius strip in half along the band, it remains in one piece, as the limerick promises.

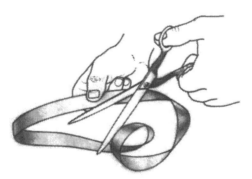

Cutting a Möbius Strip in Half

The Möbius strip has only one edge and only one side. If you run a paintbrush along the band, you'll find that when you return to where you started, you will have painted the entire surface of the band. And if you run a Magic Marker along the edge, you'll soon convince yourself that the band has only one edge.

In 1858, a scientific society in Paris offered a prize for the best essay on a mathematical subject. In the course of coming up with an essay for this competition, August Ferdinand Möbius, a mathematician in Leipzig, Germany, "discovered" the surface that now bears his name. Möbius described his discovery solely in terms of pure mathematics, with no discussion of, say, the possibility that Möbius-strip molecules exist in nature.

To be sure, the possibility of Möbius-strip molecules, for example, could not have occurred to Möbius, because the science of organic chemistry was in its infancy, and almost nothing was known about even simple molecular shapes, let alone complex ones that are mathematically interesting. Möbius made his discovery at the same time that August Kekule, at the University of Bonn, announced a finding that would become the basis of organic chemistry: carbon atoms can join to form long chains.

Kekule had first contemplated carbon chains four years earlier, in a daydream on a London bus. "One fine summer evening," he recalled, "I was returning by the last omnibus, 'outside' as usual, through the deserted streets of the metropolis, which are at other times so full of life. I fell into a reverie and lo! the atoms were gamboling before my eyes. . . . I saw how, frequently, two smaller atoms united to form a pair, how a larger one embraced two smaller ones; how still larger ones kept hold of three or even four of the smaller; whilst the whole kept whirling in a giddy dance. I saw how the larger ones formed a chain. . . . I spent part of the night putting on paper at least sketches of these dream forms."

Eleven years later, in 1865, Kekule realized that the carbon chains could curl around to form rings. Again, a dream provided the inspiration. "I was sitting writing at my textbook, but the work did not progress; my thoughts were elsewhere. I turned my chair to the fire, and dozed. Again the atoms were gamboling before my eyes. This time the smaller groups kept modestly in the background. My mental eye, rendered more acute by repeated visions of this kind, could now distinguish larger structures of manifold conformations; long rows, sometimes more closely fitted together; all twisting and turning in snakelike motion. But look! What was that? One of the snakes had seized hold of its own tail, and the form whirled mockingly before my eyes. As if by a flash of lightning I woke. . . . I spent the rest of the night working out the consequences of the hypothesis."

First, Kekule worked out the consequences for the structure of benzene, which was known to be composed of six carbon atoms and six hydrogen atoms. The six carbons formed a hexagon, Kekule concluded, with a hydrogen atom linked to each carbon.

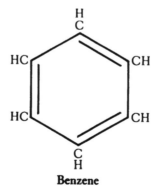

Benzene

In the 120 years since Kekule identified the shape of benzene, organic chemists have, of course, discovered the shapes of more complex molecules, such as double-helical DNAs. But

DNA

it is only recently that chemists have observed a molecule in
the shape of a Möbius strip.

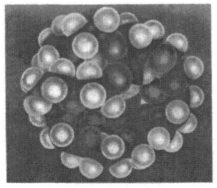

Möbius Molecule

The Möbius molecule was not found in nature but was synthesized in the laboratory by David Walba and his colleagues. He begins the synthesis with a molecule shaped like a ladder with three rungs. (Each rung is actually a carbon-

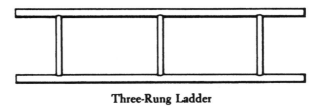

Three-Rung Ladder

carbon double bond, but we can ignore this.) The ladder is then bent around, in effect, and the ends are joined to form a loop.

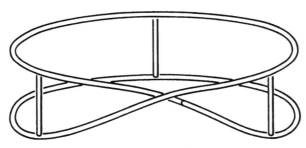

Three-Rung Möbius Strip

Half of the time the loop will simply be a circular band, but the other half of the time the loop will be a Möbius strip because, as the ends are joined, the ladder gets a half-twist.

The molecular Möbius strip shares many of the weird properties of its paper cousin. If all three carbon double bonds were broken, the molecule would remain in one piece. Breaking the bonds would be equivalent to dividing a paper Möbius strip in half along a line around its middle. For both the molecule and the paper, the result would be a single band with twice the circumference of the original.

Chemists have long known that two compounds may have the same molecular formula (that is, be composed of the same chemical constituents in exactly the same proportions) but exist as distinct chemical entities. This can happen if the chemical constituents bond to each other in different ways or at different angles. Yet, even with identical bonds, two compounds with the same molecular formula can still differ chemically. How is this possible?

The explanation lies in the branch of mathematics known as topology, the study of the properties of an object that remain the same when the object is continuously deformed. Imagine that the object is made of flexible rubber. Topologists want to know what properties remain invariant when the object is pushed and pulled but never punctured or torn. This abstract idea can be fleshed out with a specific example: the Möbius strip. Suppose you had a rubber Möbius strip that you stretched every which way. No matter how much you deformed it, the resulting shape would always have one side. Hence the property of being one-sided is of concern to topologists. When one shape can be continuously deformed into another, the two shapes are said to be topologically equivalent. Thus, the Möbius strip is topologically equivalent to whatever shape we stretch it into.

Consider now two Möbius strips, one made by twisting a strip of rubber in one direction and the other made by twisting the rubber in the opposite direction.

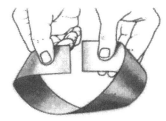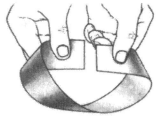

Two Kinds of Möbius Strips

Are these two Möbius strips topologically equivalent? They are not. Neither can be deformed into the other. If you looked at either strip in the mirror, the reflection would look like the other strip; the two strips are mirror images of each other.

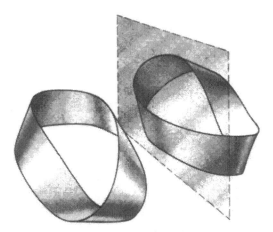

Mirror Image Möbius Strips

I must pause to issue a disclaimer to ward off nasty letters from mathematicians. Queer birds that they are, topologists do not restrict themselves to three dimensions. In four dimensions, it turns out that mirror-image Möbius strips can be deformed into each other. But I will stick to three dimensions, because molecular shapes, the ultimate subject of our inquiry, are always observed in three dimensions. And I repeat, in three dimensions, mirror-image Möbius strips are topologically distinct.

The possibility of topologically distinct mirror images is the key to how two chemical compounds with identical constituents and identical bonds can still be distinct entities.

Because the right hand and the left hand are familiar mirror images, it has been customary to refer to an object that is distinct from its mirror image as being either right-handed or left-handed. Which image in a mirror-image pair is called which is a matter of convention, just as the right side of the street is not an absolute location but depends on whether you're going up the street or down it. The two kinds of Möbius strips are referred to as right-handed and left-handed, but don't worry about which is which. Molecules that exist in right-handed and left-handed forms are said to be chiral, from *cheir*, the Greek word for "hand."

The right-handed and left-handed Möbius strips are examples of mirror-image shapes that are topologically distinct. Mirror-image shapes also exist that are topologically equivalent. To take a simple case, a circle is its own mirror image, and obviously a circle is topologically equivalent to itself.

Another example is the letter R and its mirror image, Я. Picture the R as made of soft rubber. We could transform it into its mirror image by a process of topological deformation.

Molecules, however, are not made of soft rubber. Physical restraints prevent them from being deformed any which way. Nevertheless, an R-shaped molecule can be converted into its mirror image without being bent out of shape—indeed, without being bent at all. This time, picture the R and its mirror image, Я, as stiff plastic letters on a tabletop. You can transform either one into the other simply by picking it up and turning it over.

This kind of transformation is called a rigid transformation because the object always retains its rigidity.

Many an organic molecule is rigidly chiral: it is rigidly distinct from its mirror image. The human body has a decided preference for chiral molecules of a specific handedness. Most proteins, for example, are made up of left-handed amino acids and right-handed sugars. When a chiral molecule is synthesized in the human body, only those with the desired handedness are created.

But when a chiral molecule such as a drug is synthesized in the laboratory by a nonbiological process, the result is a fifty-fifty mix of the right-handed and left-handed forms. It is difficult to remove the undesired form, so when the drug is administered, the patient receives a mix. In general, the undesired form is biologically inert and passes through without effect. But occasionally, it is harmful. This was the case with thalidomide, given to pregnant women in the early 1960s. The right-handed molecules had the desired sedating property, but the left-handed caused birth defects.

Writing in the British weekly *New Scientist*, Stephen Mason, a chemistry professor at King's College, London, noted that of the 486 synthetically produced chiral drugs that are listed in a standard pharmaceutical directory, only 88 consist of molecules of the desired handedness. The remaining 398 are fifty-fifty mixtures. "Yet," Mason concludes, "they are used in an environment (the human body) with a distinct preference for one hand. What are the effects?"

When an organic chemist analyzes a new molecule, one of the first things he tries to determine is whether it is rigidly chiral—rigidly distinct from its mirror image. Here topology may help. If the molecule is topologically distinct from its mirror image, then it's rigidly distinct, too, since a rigid transformation is but one of the many encompassed by topology. Take our old friends R and its mirror image, Я. In deforming one into the other, we achieved an intermediate shape Я that has reflection symmetry: its left half is a mirror image of its right half.

Topologists know that if a shape can be deformed into something that has reflection symmetry, the shape itself can be deformed into its mirror image. This means that a chemist can rule out the chirality of a molecule if the molecule can attain a shape that has reflection symmetry.

This insight often proves useful. Walba had synthesized the molecular Möbius strip from a molecule in the shape of a three-rung ladder. He asked me to visualize a similar synthesis from a two-rung ladder. Is the resulting shape chiral? As the following diagram shows, it is not, because it can be transformed into a shape that has reflection symmetry.

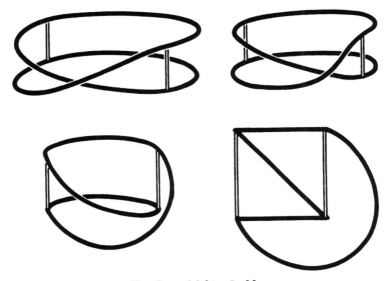

Two-Rung Möbius Ladder

Unfortunately, the three-rung Möbius molecule seems to be immune to this kind of analysis. After many thought experiments, Walba conjectured that it cannot be deformed into a shape that has reflection symmetry. If a deformation had exhibited reflection symmetry, he would have concluded that the three-rung Möbius shape can be deformed into its mirror image. But is the converse true? Does any deformation's failure to exhibit reflection symmetry mean that the molecule itself cannot be deformed into its mirror image?

Pitfalls lie in the way of an easy answer. Walba asked me to consider two rubber gloves, one right-handed and the other left-handed.

Two Gloves

The gloves are obviously mirror images, but are they topologically equivalent? Certainly, the gloves are not rigidly equivalent, because turning one of them over, as we did to the letter R, gets us nowhere. We can make the gloves equivalent, however, by turning either one of them inside out!

Turning a Glove Inside Out

(The topologist therefore finds himself in the unique position of considering gloves to be neither right-handed nor left-handed.) At no step in the turning-inside-out process did a glove have reflection symmetry.

We might be tempted to conclude that the glove is a counterexample: a shape that is topologically equivalent to its mirror image but none of whose deformations has reflection symmetry. This conclusion would be wrong. We simply have not deformed the glove enough. If we struggled with the glove, we could, at least in theory, deform it into the shape of a circular disk, which has reflection symmetry (along any diameter).

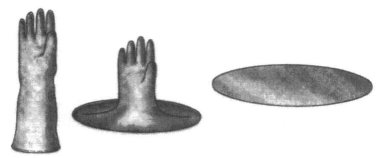

Turning a Glove into a Disk

The upshot is that Walba's innocent investigations in matters chemical have posed a big question for topologists: If no possible deformation of a shape exhibits reflection symmetry, does it follow that the shape itself is not topologically equivalent to its mirror image? This question is a fundamental one, but it does not seem to have been addressed in the mathematical literature.

All this bears on an important philosophical question: More often than not, do new ideas in the physical sciences inspire new ideas in mathematics, or is it the other way around? To put it another way, which comes first—science or mathematics? Many philosophers have confronted this question, but their answers seem to be as unsatisfactory as those given to the old question about the chicken and the egg.

In both cases, the conclusion one reaches seems to be a matter not of indisputable evidence but of teleological taste. High-handed mathematicians who follow in the footsteps of Plato claim that their discipline is divorced from physical reality. Numbers, they say, would exist even if there were no objects we could count. Mathematicians who are less dogmatic may concede that science and math are intimately connected, but they maintain that math comes first. As evidence, they point to group theory, a branch of mathematics that originally, in the 1830s, was totally without physical applications but that has recently been used by particle physicists to bring order to the study of the multitude of subatomic particles discovered in the past two decades.

But physicists who believe in the priority of their own discipline also have history on their side. Isaac Newton, for example, invented the now famous branch of mathematics called calculus because he needed a mathematical tool with which to analyze exceedingly small intervals of space and time. I think the only fair generalization to make, although it is neither exciting nor particularly informative, is that math and science each profits from the other. The story of the Möbius strip is a good example of the intricate give-and-take between mathematics and the physical sciences. Conceived in an 1858 essay competition as a construct of pure mathematics, the Möbius strip now comes up in chemistry, and its manipulation by chemists has in turn raised questions for pure mathematicians.

You may be amused to know that the Möbius strip is of service not only to chemists but to industrialists as well. The B. F. Goodrich Company has a patent on a Möbius-strip conveyor belt. In an ordinary conveyor belt, one side is subject to more wear and tear. In a Mobius belt, however, the stress is spread out over "both sides," so that the belt lasts twice as long.

7

THE CASE OF THE MISSING
THREE-HOLED HOLLOW SPHERE WITH
ONE HANDLE

During the 1940s and 1950s, many of the keenest minds in mathematics worked passionately to develop the first electronic computer. They were successful, of course, and, in the past three decades, the electronic brainchild of mathematicians has revolutionized many areas of science but not, ironically, mathematics itself. "Look around this department," says the Stanford mathematician Joseph Keller. "We have fewer computers than any other department on campus—and that includes French literature."

"It's a funny thing," says Robert Osserman, a colleague of Keller's who's been at Stanford for thirty years. "The absence of computers is clearly a combination of conservatism on the part of mathematicians—their not wanting to take time out to really learn how to use the computer effectively—and a strong conviction that a lot of the time when you use a computer, it's just an excuse for not thinking harder."

These days, however, Keller and Osserman are more sanguine about the future of the computer in mathematics, thanks to a spectacular discovery by a former Stanford student, David Hoffman, who is now at the University of Massachusetts at

Amherst. With the aid of an innovative computer graphics system, Hoffman and a fellow geometer, William Meeks III of Rice University, discovered an infinite number of graceful surfaces that adhere to certain strict criteria that only three surfaces had been known to meet. These strange new surfaces make the Möbius strip seem mundane and ordinary. They certainly fill a gap in mathematics, but, like the Möbius strip, they may also prove to be useful outside of pure mathematics, in disciplines as diverse as embryology and dentistry.

The computer's most famous contribution to fundamental mathematics is a ten-year-old result that disturbed the old guard. In 1976, Kenneth Appel and Wolfgang Haken of the University of Illinois proved the celebrated four-color-map theorem, which states that at most four colors are needed to paint any conceivable flat map of imaginary countries in such a way that no two bordering countries have the same color.

I was an undergraduate at Harvard at the time, and when word of the proof reached Cambridge, my instructor in differential equations cut short his lecture and uncorked champagne. For 124 years, the four-color-map theorem—so very seductive in its simple wording—managed to confound a parade of distinguished mathematicians and dedicated amateurs, who searched in vain for a proof (or, conceivably, a counterexample). Glasses held high, my tweedy classmates and I followed our instructor's lead in toasting Appel and Haken for having conquered this mathematical Mount Everest.

Some days later, we learned that Appel and Haken's proof made unprecedented use of high-speed computers: 1,200 hours logged among the three of them. The proof is simply too long to be checked by hand. (The curious reader who has a decade to kill can study more than 460 pages of checklists in the *Illinois Journal of Mathematics*, volume 21.)

I can remember how upset we were. The proof did not fit in with the vision of mathematics championed by Paul Erdös, an itinerant septuagenarian who is the world's most prolific mathematician. Erdös believes that God has a thin little book that contains short, elegant proofs of all significant mathematical theorems. The four-color-map theorem is undoubtedly in that book, but Appel and Haken's proof certainly isn't.

Our instructors shared our dismay. Some were afraid that the computer might have slipped up and made a subtle error. Others accepted that a machine had helped to prove the theorem, but did not give up hope that someday the proverbially bright high school student would construct a short, classy proof, the one savored by Erdös's God. Still others wondered whether the tediously long proof would be the last word on the subject; they speculated that the four-color-map theorem was representative of a whole class of interesting theorems for which simple proofs did not—and could not—exist.

Today, more than a decade later, the verdict on Appel and Haken's work is still not in, but it certainly hasn't ushered in an age of computer proofs. To be sure, computers have found new prime numbers and solved Archimedes' cattle problem, but that's not proving a theorem. In fact, no famous theorem since the four-color-map theorem has been disposed of by a machine. Hoffman and Meeks used the computer in another way, which may be the way of the future. They harnessed the computer's number-crunching power to obtain an insight that enabled them to forge ahead without the machine's help and prove a fundamental result.

For 150 years, mathematicians have studied the shape of soap films, and the surfaces that Hoffman and Meeks discovered are related to these shapes. If a circular loop of wire is dipped into a soap solution and then extracted, the soap film that spans the loop has the shape of a flat disk. This shape is known as a minimal surface because, of all surfaces that could conceivably span the wire loop, the flat disk has the least area.

Flat-Disk Soap Film

If, instead, two circular loops of wire, held one on top of the other a short distance apart, are dipped into the soap solution, a film that spans both wires will have a shape, called a catenoid, that resembles that of a nuclear power plant's cooling tower.

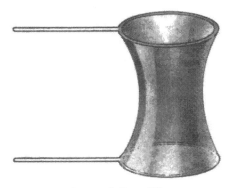

Catenoid Soap Film

It, too, is a minimal surface; no other surface that bounds both wire loops has a smaller area. Nature favors minimal surfaces because they are physically stable: minimal area means minimal stored energy.

Leave it to mathematicians to extend the concept of a minimal surface from the kitchen-physics world of soap films to the unearthly realm of the infinite. The idea of an *infinite* minimal surface may seem like a contradiction, since any surface that extends infinitely in one or more directions must have an unbounded area. When a mathematician says that an infinite surface is minimal, he means that any sufficiently small finite region of that surface minimizes area the way a soap film does. In other words, if you took a Magic Marker and drew a small enough closed curve anywhere on that infinite surface, the piece of surface within that curve would have the smallest possible area given that particular curve as a boundary.

The plane is the simplest example of an infinite minimal surface; the flat-disk soap film is just a piece of the plane. If the ends of a catenoid are extended forever, the result is another infinite minimal surface. The plane and the infinitely extended catenoid are surfaces that do not intersect themselves. They do not double back on themselves, not even near infinity.

Surfaces such as the plane and the unbounded catenoid can be deformed into a simple finite object: a hollow sphere that has some number of tiny holes and some number of hollow handles. (Picture, if you will, a hollow handle on a piece of luggage that allows air in the luggage to flow through the handle back into the bag. Mathematically, each handle serves to increase the "connectivity" of the surface, because a cut through the handle will not divide the surface into pieces.)

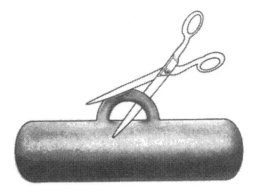

Luggage

Mathematicians, with their hyperactive imaginations, pretend that surfaces are made of superflexible rubber. If one of these surfaces can be deformed into another by stretching, shrinking, twisting, or any other manipulation that does not involve ripping, puncturing, or filling holes, the two surfaces are said to have the same topology.

A hollow sphere, for example, can be stretched into an egg-shaped surface, and so the two have the same topology.

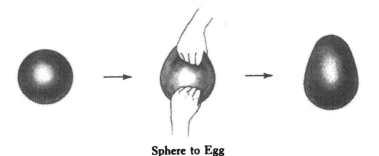

Sphere to Egg

The plane is topologically the same as a sphere pierced by a single, tiny hole, because the hole can be pulled open indefinitely in this peculiar world that would make Charles Goodyear weep.

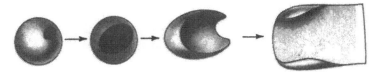

One-Holed Sphere to Plane

The catenoid has the same topology as a hollow sphere that has two holes in it; each hole can be widened and stretched to infinity. (In general, each hole in a multiholed hollow sphere can be extended to infinity.)

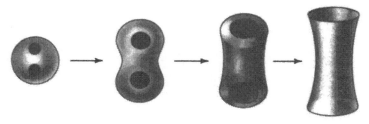

Two-Holed Sphere to Catenoid

When Hoffman and Meeks began their investigation, mathematicians knew of only one other infinite minimal surface, besides the plane and the unbounded catenoid, that did not intersect itself and that could be modeled, by a rubber-sheet deformation, on a holey hollow sphere (with or without handles). This surface is the unbounded helicoid, which resembles a screw extended to infinity. Like the plane, the helicoid has the same topology as a one-holed hollow sphere.

Helicoid

These three minimal surfaces had been known for almost two hundred years, and a string of results in the past decade made it seem unlikely that a fourth kind existed. For example, in 1981 Rick Schoen at the University of California, San Diego, proved that a hollow sphere with two holes could serve as a model only for the catenoid and not for any other infinite minimal surface that was free of self-intersections. In the same year, the Brazilian mathematician Luquesio Jorge proved that a hollow sphere with three, four, or five holes and no handles could not be a suitable model.

Jorge's Unsuitable Spheres

"Because the existence of new minimal surfaces was ruled out in all sorts of special cases," says Hoffman, "a lot of people believed, and tried to prove, that no new examples could exist. They weren't successful, but there was a general feeling that the reason they did not succeed was not because they were futilely trying to prove something that was actually false, but because they didn't have sophisticated enough mathematical tools."

In November 1983, Hoffman learned of a graduate student in Brazil named Celso Costa, whose dissertation included some thorny equations for a proposed surface that Costa was able to prove was infinite, minimal, and topologically the same as a three-holed hollow sphere with one handle.

Three-Holed Hollow Sphere with One Handle

But neither Costa nor anyone else knew what the proposed surface looked like, because the equations defining it seemed hopelessly complex. Moreover, no one knew whether the surface intersected itself—something it wasn't permitted to do if it was to join the hallowed ranks of the plane, the unbounded catenoid, and the unbounded helicoid.

The question of self-intersection is not an easy one. "When you have the equations for a surface," explains Hoffman, "you can't compute some quantity that says 'Yes, it intersects itself' or 'No, it doesn't.' Basically, all you can show is that a particular piece of the surface doesn't intersect another piece." But that doesn't get you very far for an infinite surface, because you would have to compare an infinite number of pieces.

Hoffman's plan was to use a computer to calculate the coordinates of the surface's core and then draw a picture of that core. Conventional computer-graphics packages, however, would not have helped, because they deal chiefly with cubes and spheres and other mundane shapes used by engineers, not esoteric mathematical surfaces that self-intersect or extend to infinity. Luckily, he learned of a University of Massachusetts graduate student, James Hoffman, who was developing a novel computer-graphics package.

"Our game plan," says David Hoffman, "was to use the computer to look at the surface. If we saw a self-intersection, we planned to publish a little paper ruling out this example. We'd probably have to publish in a lousy journal, because in mathematics it's hard to publish negative results of this sort. And if we didn't see a self-intersection, we didn't know what we'd do, except work very hard to prove that the surface was free of self-intersection."

The computer-generated picture, however, defied their expectation. Not only was it free of self-intersection; it was highly

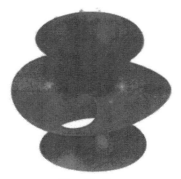

New Surface

symmetric as well. It contained two straight lines that met at right angles. After "viewing" the core of the surface from different angles and thinking long and hard, Hoffman realized that the surface could be decomposed into eight identical pieces.

In physics, seeing is believing; in mathematics, that's not enough. But having seen the symmetries, Hoffman and Meeks put aside the picture and were able to prove just from the equations that the surface did not intersect itself. They had discovered, to their amazement, a fourth infinite minimal surface, made from two catenoids and a plane that all sprout from a Swiss-cheese core. Three months later, they were able to demonstrate the existence of infinitely many such surfaces,

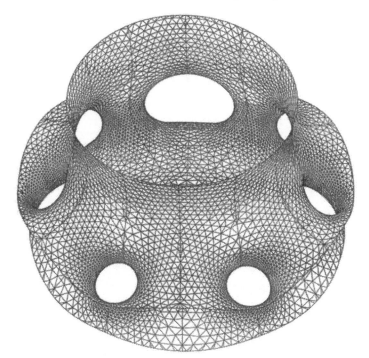

Another New Surface

each topologically equivalent to a three-holed hollow sphere
with some number of handles.

New Spherical Models

After Hoffman and Meeks published pictures of the core of
the first new surface, they were contacted by a biologist at the
University of Cambridge who thinks that developing embryos
can assume that shape. Area-minimizing surfaces often occur
naturally in interfaces between organic and inorganic materials, because such surfaces minimize surface tension. A New
York dental surgeon called Hoffman up and said the picture
looked exactly like what should be used for bone implants to
which false teeth could be secured. He thought, says Hoffman,
that "a minimal surface would be less destructive because it
would have less contact with the bone. Plus, there would be a
lot of 'handles' for bone to go through."

Even if these real-world uses of the surface do not pan out,
Hoffman and Meek's discovery is a monumental one. It ex-
poses the poverty of recent wisdom about infinite minimal
surfaces. And it is testimony to the utility of the computer in
pure mathematical research. It is hard to take issue with a
staggering, computer-aided advance on a problem that defied
understanding for almost two centuries.

MACHINES

The computer has proved useful to mathematicians in finding large primes, solving Archimedes' cattle problem, breaking codes, proving the four-color-map theorem, and discovering new shapes. Nevertheless, there are subtle limits to what the computer can do.

Since the 1930s, mathematics has faced a revolution as fundamental as the two revolutions—general relativity and quantum mechanics—that shook the foundation of physics and toppled classical theories about space, time, and causality. The landscape of mathematics has been radically transformed by what Morris Kline of New York University calls "the loss of certainty." A new breed of work has focused not on the power of mathematical computation but on its limitations. Meaningful computational problems have been identified that either cannot be solved in principle or can be solved in principle but not in practice.

The classic example of a meaningful problem that cannot be solved in principle is the "halting problem," posed in 1936 by Alan Mathison Turing. Turing considered the problem of whether a computer program will sooner or later come up with a result and halt. The halting problem is not the exclusive concern of armchair theorists. It can easily arise in practice.

"You can imagine," says Michael Sipser, an MIT theoretician of computer science, "wanting to know the answer, especially in the old days, when you wrote your program on cards and then submitted them to the computer center. They'd run them overnight and get back to you the next day. And you'd have an account with, say, a hundred bucks in it. Every once in a while, the program would have an infinite loop and burn up gobs of money. You'd get nothing out of the program, since it was stuck in an infinite loop. Either your account would run out of money or somehow the machine would notice that it had been going for a very long time and shut itself off.

"So, gee, you'd think, why not test the program first. And if it has an infinite loop in it, don't run it." But, amazingly, this natural idea cannot be carried out, because Turing proved that no test is possible that will work for all programs.

Besides Turing's proof of the impossibility of solving the halting problem, the year 1936 witnessed another assault on the illusory goal of absolute mathematical knowledge. The logician Alonzo Church proved that the so-called decision problem was unsolvable: there can never be a general procedure for deciding whether a given statement expresses an arithmetic truth. In other words, no computer will ever exist that can spew out all the truths of arithmetic. Nor for that matter will a machine ever be able to determine the truth of every conceivable arithmetic statement you give it. There simply is no recipe for finding arithmetic truth.

In recent years, the attention of the mathematics community has turned from problems that cannot be solved in theory to problems that can be solved in theory but not in practice. Among the most notorious of these are the ones that IBM's Larry Stockmeyer calls "intrinsically difficult"—a euphemism, if there ever was one. He asks you to conceive of the most powerful computer imaginable. This ideal computer would be the size of the entire universe (perhaps 100 billion light-years in diameter). It would be built from hardware the size of the proton (10^{-13} centimeter in diameter), through which signals would race at the speed of light (3×10^{10} centimeters per second). It would have the ability to work on a single problem for twenty billion years, which is longer than the estimated age of the universe. An intrinsically difficult problem has the mind-boggling distinction of being solvable in principle but not by the most powerful computer imaginable running for the age of the universe.

One such problem involves playing chess not on the usual eight-by-eight board but on an n-by-n board (where n is an arbitrarily large number) with an unlimited number of pieces (except that there can be only one king on each side). We want a program for determining whether, in any given position, one of the players, say White, has a forced win. One program that works in principle but not in practice would consider all possible moves for White, then all possible responses for Black, then all counterresponses for White, and so on, until all conceivable continuations had been examined through to the end.

The drawback of this exhaustive-search program is that it is too slow: there are so many possible continuations that even the ideal computer could not look at all of them in twenty billion years. In 1981, David Lichtenstein at Yale and Aviezri Fraenkel, an Israeli mathematician, proved that for sufficiently large boards there is no faster program. In other words, there are no shortcuts to the time-consuming exhaustive search. This chess problem will always defy computer analysis, even though we know it has a solution.

In the next four chapters, we'll look at the power and limitations of computing, both in theory and in practice.

8

TURING'S UNIVERSAL MACHINE

In February 1952, Alan Mathison Turing—mathematician *extraordinaire*, pioneer in computer science, a key force in breaking the Nazis' famous Enigma cipher—was arrested in Manchester, England, for the crime of "Gross Indecency contrary to Section 11 of the Criminal Law Amendment Act 1885." Turing's home had recently been burgled by an acquaintance of Arnold Murray, a nineteen-year-old unemployed youth with whom Turing had had sexual relations. When Turing reported the burglary to the police, he told them about his affair with Murray, naively believing that a royal commission was about to legalize homosexuality. Two months later, Turing was tried for six sexual offenses, was convicted of all six, and, instead of being sent to prison, was given a year's probation on the condition that he undergo "organo-therapic treatment," a program of regular doses of androgynizing female hormones. On June 7, 1954, at the age of forty-one, Turing took his own life by eating half of an apple he had dipped in a cyanide solution.*

In the field of artificial intelligence, Turing has achieved

*The tragic details of Turing's last years are spelled out in Andrew Hodges's sympathetic biography, *Alan Turing: The Enigma* (New York: Simon & Schuster, 1983).

immortality in connection with two fundamental concepts: the Turing test and the Turing machine. The Turing test was his idea of how to determine whether a machine can think. The test calls for the machine and a randomly chosen person to be separated from an interrogator, who asks them each an unlimited number of questions through an intermediary. Turing thought that if the interrogator failed to distinguish between the machine and the human being, it meant that the machine was thinking. In other words, if the machine passes for intelligent, it *is* intelligent.

An explanation of the idea of the Turing machine requires some background. At Cambridge University, where Turing was made a fellow of King's College in 1935, he was exposed to the revolutionary developments in physics that toppled traditional ideas of causality and determinism. According to the Newtonian worldview, if sufficient information is known about a physical system, its entire future can be predicted.

In 1795, Pierre-Simon Laplace, a French mathematician and an avid Newtonian, put it this way: "Given for one instant an intelligence which could comprehend all the forces by which nature is animated and the respective situations of the beings who compose it—an intelligence sufficiently vast to submit these data to analysis—it would embrace in the same formula the movements of the greatest bodies and those of the lightest atom; for it, nothing would be uncertain and the future, as the past, would be present to its eyes."

The introduction of quantum mechanics, however, in the early part of this century put an end to the idea that the future is completely determined by the present and the past. Cambridge University was in the 1930s at the center of the philosophical havoc engendered by quantum mechanics, particularly by the principle that the observer always influences the observed. Turing found this idea unsettling, and he was drawn

to mathematics because it seemed to deal with absolute entities, independent of observers. As G.H. Hardy, the Cambridge
number theorist, put it, "317 is a prime, not because we think
so, or because our minds are shaped in one way rather than
another, but *because it is so*, because mathematical reality is
built that way."

Turing set himself the task of answering a difficult question
that cut to the heart of the nature of mathematical reality: Is
there a mechanical way of determining whether any given
statement in mathematics is true or not? To answer the question, he came up with the notion of a "universal machine"—
the Turing machine—that could routinely answer mathematical questions. By introducing the idea of a machine that could
do mathematics, Turing aimed to bolster the status of mathematics as a subject independent of human affairs. Ironically,
however, Turing found that some mathematical questions—
involving, for example, the generation of numbers that are
nonrepeating decimals—could not be solved by mechanical
means by machine or man.

The Turing machine is an extraordinary concept. From the
point of view of the range of its behavior, the machine is very
limited. Indeed, it is so limited that even if you know nothing
about computer programming (and if the whole subject scares
you), in no time at all you'll understand the Turing machine's
"inner" workings and be gleefully writing programs for it.
From a computational point of view, however, it can do anything—or, I should say, anything that a human mathematician
can do and anything that the most powerful computer imaginable could do. And if that's not remarkable enough, let me
cryptically add that the Turing machine, in spite of its name,
need not be a machine. It could be a person or group of people.

So what are the elements of this Turing machine? Well, first
there is a long tape—imagine a narrow strip of paper on which

vertical lines have been drawn that divide the strip into square cells.

If a given cell is not blank, it contains one symbol from a finite alphabet of symbols. The Turing machine has the ability to scan the tape one cell at a time, generally beginning with the leftmost cell that contains a symbol. If a scanned cell is empty, the machine can leave it empty or print a symbol in it. If the scanned cell contains a symbol, the machine can leave the symbol unaltered, erase the symbol and print another one in its place, or erase the symbol and leave the cell empty. Then the machine stops, or it scans the cell immediately to the left or immediately to the right.

What the machine does to a scanned cell and which of the two adjoining cells it scans next depends on the state, or internal configuration, of the machine. The number of states, like the number of symbols, must be finite. A machine state is like a mental state—it's what's in the machine's "mind." One need not be more precise (or more metaphysical) about the nature of a state in order to understand the operation of a Turing machine. The machine is defined by a "table of behavior," which stipulates what the machine will do for each possible combination of symbol and state.

A specific example—that of adding two numbers—will do much to clarify these abstract ideas. Suppose we write the numbers in "unary" notation, in which an integer n is represented by a string of n *s one * per cell in n consecutive cells. On this basis, * * represents 2 and * * * * * represents 5. The advantage of unary notation is that only one symbol, *, not ten different digits, is needed to represent any given positive integer. To add 2 and 5, * * and * * * * * are printed on the tape, with a blank cell between them so that the two strings of *s can be distinguished.

Addition

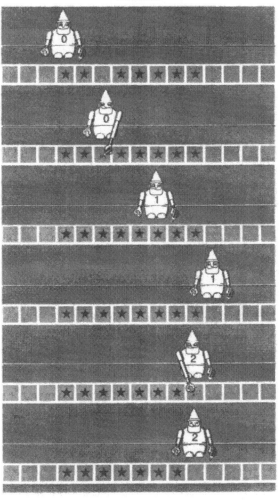

ADDITION

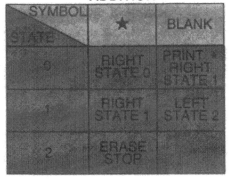

SYMBOL / STATE	*	BLANK
0	RIGHT STATE 0	PRINT * RIGHT STATE 1
1	RIGHT STATE 1	LEFT STATE 2
2	ERASE STOP	

The table of behavior with the diagram explains how the machine goes about adding the two numbers, but before I discuss the specifics of the table, I want to describe the additive process in general terms. The clever machine finds the blank cell between the numbers, prints * in it (thus leaving a string of eight *s on the tape), goes to the end of the string of *s, and erases the last *. This leaves * * * * * * *, which, in unary notation, is 7, the sum of 2 and 5.

Let's look at the table now. The machine states are always listed in the leftmost column. In this case, there are three states, numbered respectively 0, 1, and 2. The symbols (and the word *blank,* corresponding to an empty cell) are always listed across the top of the table. Here the only symbol is *.

The machine begins in state 0 and, according to convention, scans the leftmost symbol on the tape (in other words, the first * in * *). The table describes what the machine does for the combination of state 0 and symbol *. The machine leaves the symbol unaltered, switches to the next cell to the right, and stays in state 0. What does the machine do in this cell? Since the machine is still in state 0 and since the symbol in the cell is still a *, it does the same thing as before: it leaves the symbol alone, switches to the next cell to the right, and stays in state 0.

Now, for a change, the cell is blank. The combination in the table of state 0 and a blank symbol tells the machine what to do. It prints *, switches to the cell on the right, and enters state 1. A * is in this cell, and so the combination of state 1 and symbol * describes the machine's behavior: switch to the next cell to the right and stay in state 1. This step is repeated four times because a * keeps popping up. When the machine reaches the blank cell at the end of the string of five *s, it moves back one cell to the left and enters state 2. In the cell is a *, which the machine erases. Then the machine stops and rests on its laurels.

The power of this approach is that the same table of behavior can generate the sum of any two numbers, no matter what their size, written in unary notation with a single blank cell between them. In state 0, the machine simply scans the first number cell by cell until it reaches the blank cell, in which it prints a *. In state 1, it scans the second number cell by cell until it reaches a blank cell, turns around, and comes to rest on the last *. In state 2, it simply erases this *. And, presto, the answer is on the tape.

This method of addition is known as a *finistic* approach, because the table of behavior includes a finite number of states and a finite number of symbols. Yet this finistic approach can generate an infinite range of numbers. For the Turing machine to be able to handle any conceivable sum, the paper tape must be unlimited in length; if it were, say, only 1,000 cells long, it could not work with a number greater than 1,000.

When the Turing machine finishes adding the two numbers by this method, the tape will contain only the answers and not the original numbers. It is tempting to try to write a table of behavior that preserves the original numbers. One way that immediately comes to mind is to have the machine "count" the eight *s in the two strings of *s. Surprisingly, however, the Turing machine cannot count. Suppose that when it scans the first *, it kicks into state 1. Each time it scans another *, it kicks into the next state. Thus, after scanning the fifth *, the machine is in state 5, and after scanning the twenty-third *, it's in state 23. It would seem that by this method the Turing machine can count the number of *s; when it has scanned them all, the number of the state it is in will correspond to the number of *s. Nevertheless, this approach is flawed. Do you see why?

The problem is that the method is not finistic. It requires an infinite number of states. If there were, say, only 5 states, it could not count more than five *s, and so would be restricted to sums of 5 or less. If there were 50,000 states, it could not handle more than 50,000 *s. In other words, for a finite number of states n, it could not handle more than n *s. And this will not do, because we are looking for a method that applies to any addition problem whatsoever.

If an infinite number of states (or an infinite number of symbols) were allowed, the table of behavior could not be written out. The requirement that the method be finistic implies that the table of behavior can be written out and, hence, followed in a routine, mechanical way.

It is time to take up the intriguing idea that a Turing machine need not be a machine. It is the table of behavior—the software, if you will—that defines the Turing machine. Any entity, be it a computer, a human being, a mermaid, a disembodied spirit, or the Kremlin, is a Turing machine if it follows a table of behavior. *You* are a Turing machine if you add two numbers on a tape according to the table of behavior in the addition diagram. In a brilliant paper, Turing was able to demonstrate that no mathematician and no computer, either in theory or in practice, could do something that his Turing machine could not do. A supercomputer may be able to solve problems much faster, but the slowpoke Turing machine is also able to solve them.

The best way to grasp the essence of a Turing machine, and the power of finistic methods, is to write tables of behavior yourself. I challenge you to write a table for a Turing machine that can subtract numbers in unary notation. Be warned that you must build into the table a way of letting the machine know that it has completed the computation. Otherwise, since the tape may be of any length, it might continue scanning blank cells forever. The diagram shows a table of behavior for subtraction. Other tables will also work.

Subtraction

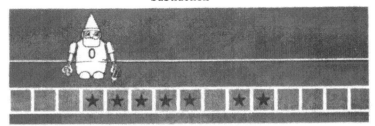

SUBTRACTION

SYMBOL STATE	★	BLANK
0	RIGHT STATE 0	RIGHT STATE 1
1	RIGHT STATE 2	RIGHT STATE 1
2	LEFT STATE 3	LEFT STATE 5
3	ERASE LEFT STATE 4	STOP
4	ERASE RIGHT STATE 1	LEFT STATE 4
5	ERASE LEFT STATE 6	
6	ERASE STOP	LEFT STATE 6

The second problem is to write a table of behavior for a machine that can test whether a sequence of Ps and Qs written on its tape is a palindrome—a sequence that reads the same forward as backward. One approach is to have the machine compare the first symbol with the last symbol, the second symbol with the second-to-last symbol, and so on. But, remember, the approach must be finistic. If the sequence is a palindrome, you can have the machine print a Y, and if it's not, you can have it print an N. One such approach, adopted from an article that Andrew Hodges wrote for the British weekly *New Scientist,* is shown.

The drawback of Hodges's approach is that, although the table has only six states, the machine wastes some time shifting back and forth along the string of symbols. Time can be saved

Hodges's Palindrome Check

PALINDROME CHECK

SYMBOL STATE	P	Q	BLANK
0	ERASE RIGHT STATE 1	ERASE RIGHT STATE 2	PRINT Y STOP
1	RIGHT STATE 1	RIGHT STATE 1	LEFT STATE 3
2	RIGHT STATE 2	RIGHT STATE 2	LEFT STATE 4
3	ERASE LEFT STATE 5	ERASE PRINT N STOP	
4	ERASE PRINT N STOP	ERASE LEFT STATE 5	
5	LEFT STATE 5	LEFT STATE 5	RIGHT STATE 0

if the machine, after comparing the first symbol with the last symbol, compares the second-to-last symbol with the second symbol (rather than the other way around), then the third symbol with the third-to-last symbol, then the fourth-to-last symbol with the fourth symbol, and so on. Such an approach, whose table of behavior is shown, requires ten states. The longer program is the price paid for shortening the computing time.

Ten-State Palindrome Check

SYMBOL STATE	P	Q	BLANK
0	ERASE RIGHT STATE 1	ERASE RIGHT STATE 2	PRINT Y STOP
1	RIGHT STATE 1	RIGHT STATE 1	LEFT STATE 3
2	RIGHT STATE 2	RIGHT STATE 2	LEFT STATE 4
3	ERASE LEFT STATE 5	PRINT N STOP	
4	PRINT N STOP	ERASE LEFT STATE 5	
5	ERASE LEFT STATE 6	ERASE LEFT STATE 7	PRINT Y STOP
6	LEFT STATE 6	LEFT STATE 6	RIGHT STATE 8
7	LEFT STATE 7	LEFT STATE 7	RIGHT STATE 9
8	ERASE RIGHT STATE 0	PRINT N STOP	
9	PRINT N STOP	ERASE RIGHT STATE 0	

The last challenge is to write a table of behavior for a Turing machine that can test whether two strings of Ps and Qs, separated by a blank cell, are anagrams. Again, Y stands for "yes" and N stands for "no." Here's a hint: the machine prints a dummy letter, R, while solving the problem. One possible answer appears right.

Anagram Check

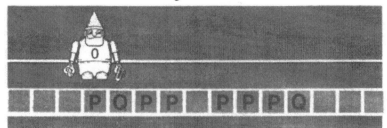

ANAGRAM CHECK

SYMBOL STATE	P	Q	R	BLANK
0	ERASE PRINT R RIGHT STATE 1	ERASE PRINT R RIGHT STATE 2		PRINT Y STOP
1	RIGHT STATE 1	RIGHT STATE 1	RIGHT STATE 1	RIGHT STATE 3
2	RIGHT STATE 2	RIGHT STATE 2	RIGHT STATE 2	RIGHT STATE 4
3	ERASE PRINT R LEFT STATE 5	RIGHT STATE 3	RIGHT STATE 3	PRINT N STOP
4	RIGHT STATE 4	ERASE PRINT R LEFT STATE 5	RIGHT STATE 4	PRINT N STOP
5	LEFT STATE 5	LEFT STATE 5	LEFT STATE 5	LEFT STATE 6
6	LEFT STATE 6	LEFT STATE 6	RIGHT STATE 0	

9

DID WILLY LOMAN DIE IN VAIN?

It may be deflating to learn that in a certain fundamental sense computers and mathematicians are nothing other than Turing machines in disguise. On the other hand, it could be seen as encouraging, since the seemingly simplistic Turing machine turns out to be able to solve all sorts of computational problems. The theoretical resemblance between mathematician and machine holds not only for the problems they can solve but also for those they cannot solve.

Every day in industry, computers routinely tackle computational problems that are too time-consuming to be solved by any known method. Industry, however, needs solutions to these problems, and so the computers—with the complicity of their programmers—often spew out answers that are less than optimal. Many of these problems are forms of the notorious traveling-salesman dilemma: given a network of cities and roads, find the shortest round-trip route that takes a salesman to each city exactly once. The only known algorithm—a sure-fire, step-by-step recipe—for solving a traveling-salesman problem is the laborious, uninsightful one of trying every possibility.

Mathematics, it seems, offers Willy Loman no relief.

For the past fifteen years, mathematicians have wondered whether their failure to find a clever, faster algorithm reflects their ignorance or the inherent difficulty of the problem itself. According to current wisdom, there is no faster algorithm, not even in theory, but no one has yet been able to prove this. The search for a proof is the hottest pursuit in theoretical computer science, and mathematicians who work in this area are known as complexity theorists.

When mathematicians speak of a guaranteed method of solution, they mean an algorithm. Don't be put off by the formidable sound of this word, which is a corruption of the last name of a ninth-century Persian mathematician, Abu Ja'far Mohammed ibn Musa al-Khowarizmi, whose semantic legacy also includes the word *algebra*. The bark of an algorithm is much worse than its bite. You already have an intuitive notion of what an algorithm is.

Remember when in grade school your English teacher had you write excruciatingly complete instructions for carrying out some boring, routine task, like tying a shoelace? Then the teacher would ask Johnny Wiseguy to tie his shoelace by following your instructions exactly to the letter. (The really mean schoolmarms would have you read the instructions aloud as Johnny Wiseguy did his thing.) Of course, he would immediately fail—and make a big deal of it—because you left out some basic step that seemed second nature to you, like grabbing the lace by its plastic-encrusted end rather than in the middle. If you had managed to write out the instructions in unambiguous detail, you would have an algorithm for shoelace tying. An algorithm is simply a step-by-step procedure in which everything is explicitly stated so that a problem can be mechanically solved. Every step must be laboriously spelled out, nothing being left to chance, intuition, experience, interpretation, or imagination.

Mathematicians, of course, are more interested in algorithms for computational problems than for shoelace tying. An algorithm for adding two whole numbers, based on the way the schoolmarms taught us to do it with paper and pencil, would make explicit such steps as lining the numbers up flush right, one above the other, drawing a line below them, doing the calculation from right to left, "carrying" a 1, and doing a host of other things we take for granted. The algorithm would include rules like "If a 2 in one number is above a 4 in the other, write a 6 below them" and "If a 3 in one number is above a 6 in another, write a 9 below them."

The power of an algorithm is that it can be applied to all instances of a problem. The addition algorithm, for example, can find the sum of any two whole numbers. What it costs you to spell out an algorithm in complete detail, you gain in having at your disposal a method of solution that is guaranteed to work. A computer program is either a single algorithm or a series of them. Without instructions telling it what to do every step of the way, a computer could no more add two numbers than it could simulate the tying of a shoelace. The role of the programmer is to come up with complete instructions—in other words, algorithms. When the programmer curses a bug in his program, he means that somewhere along the line he made a mistake spelling out an algorithm or translating the algorithm into computerspeak.

It should be emphasized that the user of an algorithm, be it a machine or a man, need never make a judgment. The use of the addition algorithm, for example, requires no conception of what a number is. To apply the algorithm, you blindly follow the rules. You don't have to know, say, that 5 comes after 4, that 7 is greater than 3, or even that you're working with a ten-digit number system. Much ink has been spilled in the philosophical literature on what the absence of judgments means in terms of a machine's ability to think, but the pursuit of such intriguing speculations would take us too far afield.

Mathematicians are not concerned with specific instances of the traveling-salesman problem. For a small set of cities and roads, it may be easy to find the solution, because there aren't that many possible routes to inspect. Even for a large network of cities and roads, you might be lucky and stumble on the optimal itinerary. When mathematicians say that the problem is unsolvable in practice, they mean that the only known methods that guarantee a solution are as inefficient as an exhaustive search of all the possibilities, a search that is too slow for even the superest of supercomputers.

The mathematical cognoscenti have a rigorous way of defining what makes an algorithm fast (and usable) or slow (and unusable). Suppose the number n is a measure of the size of a problem (for the traveling salesman, n would be a measure of the number of cities and roads). For an algorithm to be fast, the time it takes to execute the algorithm must increase no faster than a polynomial as the size of the problem increases. Polynomials are mathematical functions such as $2n$ (doubling), $3n$ (tripling), n^2 (squaring), n^3 (cubing), $3n^{10}$, and $64\, n^{100}$. Slow algorithms, such as the exhaustive-search method for solving the traveling-salesman problem, have execution times that increase exponentially—2^n, 6^n, or 12^n—as the size of the problem grows.

For small values of n (that is, for simple problems), a given polynomial function may equal or even exceed a given exponential function, but for large values of n, any exponential function will explosively overtake any polynomial function. For example, when n is 2, the polynomial function n^2 is 4, as is the exponential function 2^n. But when n is 10, n^2 is only 100, whereas 2^n has skyrocketed to 1,024. It was the certainty that exponentials will overtake polynomials that troubled Thomas Malthus when he compared the exponential growth of the human population with the polynomial growth of the food supply.

The fact that the only known way of solving a traveling-salesman problem is the exponetially slow approach of examining all conceivable itineraries means that we have, in this day and age, no real insight into such a seemingly simple problem. Complexity theorists are trying to prove the heady conjecture that we will never have any insight, no matter how hard we try, because there is no insight to be had.

Mathematicians do have insight into many problems that don't seem all that different. Consider, for example, a highway inspector who examines the network of roads on which the traveling salesman might drive. Eager to get home to his wife and children, the inspector wants to know if he can take a round-trip journey that will traverse every road once and only once. The inspector doesn't care about the cities; he just wants to cover every stretch of road and not repeat himself. The traveling salesman, on the other hand, doesn't care about the roads; he just wants to visit every city exactly once, while putting the least possible mileage on his car.

The question the inspector asked is easy to answer, thanks to the work in 1736 of Leonhard Euler (pronounced "oiler"), a twenty-nine-year-old Prussian mathematical wizard. The Prussian city of Königsberg (now the Soviet town of Kaliningrad) lay on both banks of the Pregel River and included the island of Kneiphof as well as a spit of land in the middle of a fork in the river. The four sections of the city were linked by a network of seven bridges.

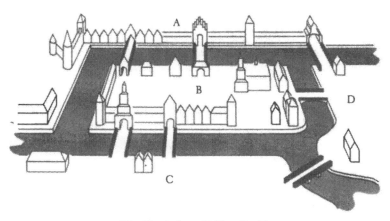

The Königsberg-Bridge Problem

Immanuel Kant, it is said, used to go on long constitutionals around the city, and the residents wondered whether it was possible to go on a round-trip stroll across all seven bridges without crossing any bridge more than once. Since the number of bridges was small, the problem was solved (in the negative) by enumerating all the possible routes—that is, in the same exhaustive, uninsightful way you would use to solve a small traveling-salesman problem.

Leave it to the prolific Euler, who fathered thirteen children and eighty volumes of mathematical results, many reportedly written down in the thirty minutes between the first and second calls to dinner, to prove in an insightful way that the journey was impossible. The spirit of mathematics cries out for an analysis of the most general case. And Euler, wanting to solve the problem not only for the people of Königsberg but

for bridge-loving strollers everywhere, tackled the general problem: "Given any configuration of the river and the branches
into which it may divide, as well as any number of bridges,
determine whether or not it is possible to cross each bridge
exactly once." If you think of the land areas as cities and the
bridges as highways, you will see that this general problem is
equivalent to the one faced by the highway inspector.

To solve the Königsberg-bridge problem, Euler represented
each bridge by a geometric line and each mass of land by a
geometric point.

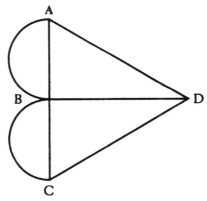

Euler's Representation of the Königsberg Bridges

In this way, he reduced the problem to its essentials, stripping
it of all extraneous information. The line-and-point representation does not distinguish, say, a wide bridge from a narrow
bridge, a particular bridge from any other bridge that also links
the same land areas, a large landmass from a tiny one, or even
an island from a river bank. However important such distinc-
tions may be in other contexts, they have no bearing on the
possibility of an exhaustive, nonrepeating stroll. That is the

beauty of a mathematical representation: it need only preserve what is relevant to the situation at hand. Free of distracting irrelevancies, the mathematician is better able to concentrate on the problem.

Euler was able to prove that a nonoverlapping stroll across all bridges could be taken only when there were either zero or two points (land areas) from which emerge an odd number of lines (bridges). A little thought supports this conclusion. If you cross a bridge to a piece of land, there must be another bridge to exit by—or else you'd be stuck there. Landmasses with an even number of bridges ensure that where there's a way in, there's another way out. Landmasses with an odd number of bridges are possible on a nonoverlapping trip, but only at the end of the journey (where you don't need a bridge out) and at the beginning (where you don't need a bridge in). Since there is only one starting point and one end point, only two land-masses can have an odd number of bridges. In Königsberg, each of the four land areas was connected to an odd number of bridges, and so a complete, nonoverlapping stroll—even without the stricter condition that it be a round-trip—was clearly impossible.

It is important to realize that Euler's conclusion about an arbitrary number of bridges and an arbitrary number of land-masses does much more than encapsulate common sense. Our reasoning merely indicated that a trip wouldn't be possible if the condition he laid down wasn't met. Euler's conclusion is much stronger (and not intuitively obvious): he proved that if this one simple condition is met (namely, that either zero or two land areas are associated with an odd number of bridges), a nonoverlapping trip is always possible.

To apply Euler's analysis to a general situation requires counting up the number of bridges for each land area. Since each bridge goes to two land areas, each bridge is counted twice. Therefore, if n is the number of bridges, Euler's analysis requires $2n$ steps. The counting of the bridges could be formulated as an algorithm, and it would be a very efficient algorithm, since the time it takes to execute it only doubles as the complexity of the problem grows. An exhaustive search of all possible itineraries, on the other hand, would exponentially explode as 2^n.

In the traveling-salesman problem, however, no shortcut is known to the inefficient exhaustive search. You can't, say, count up the number of cities connected to each road and draw some conclusion based on whether those numbers are odd or even—or, for that matter, on any other property of those numbers. Again, it's not just that we don't *know* of some property to look for. It may be that such a property doesn't exist. That's what complexity theorists are struggling to prove.

The traveling-salesman problem is not the only computational problem for which a fast algorithm eludes mathematicians. There is a whole class of problems, called NP-complete, for which the only known algorithms have execution times that balloon exponentially.* Another notorious example of an NP-complete problem is known as the clique problem: Given a large group of people, say, a hundred, is there a large number of them, say, fifty, all of whom know each other?

"You could solve this problem," says Michael Sipser, a complexity theorist at MIT, "by writing down a hundred points, one for each person, and drawing a line between points that correspond to people who know each other." Then you'd look

---

*The NP, if you must know, stands for nondeterministic polynomial, and the word *complete* means that these problems are the hardest of their kind.

for a group of fifty points all of which are connected. "It sounds like a great problem for a computer," adds Sipser, "but it isn't. The only way that we know how to solve this problem is essentially to look at all groups of fifty, of which there are a tremendous number, something like ten raised to the twenty-ninth power. To do this would take centuries, even for a fast computer."

The traveling-salesman problem, the clique problem, and all other NP-complete problems have a curious feature in common: if someone claims to have a solution to a particular instance of any one of these problems, it's an easy matter to check the solution. For the traveling-salesman problem, you would just examine the proposed itinerary and ascertain that it includes every city once. For the clique problem, you would double-check that the fifty people identified as a clique all know one another. Richard Karp, professor of computer science at the University of California, at Berkeley, likens NP-complete problems to jigsaw puzzles: "They may be hard to assemble, but when someone shows you a completed puzzle, you can tell at a glance that the puzzle has been correctly solved."

Another outstanding feature of NP-complete problems is that if any one of them can be solved by a fast algorithm, the others can, too. Moreover, it would be a trivial exercise to take a fast algorithm for one kind of NP-complete problem and alter it so that it can solve any other kind. For example, if a fast algorithm were discovered for the traveling-salesman problem, mathematicians would have at their disposal a fast method for solving the clique problem and every other NP-complete problem. Therefore, whether the traveling-salesman problem has a fast algorithm is part of the larger question of whether NP-complete problems are really as hard as they seem.

"Virtually every mathematician, I think, now believes that NP-complete problems are intrinsically hard," says AT & T's David Johnson, coauthor of the bible of this field, *Computers and Intractability: A Guide to the Theory of NP-Completeness.* "The real issue," he says, "is proving it."

It boggles the mind that mathematicians think they may be able to prove that the traveling-salesman problem and other problems of its ilk will never yield to a fast algorithm—never ever—not even under the scrutiny of the Einsteins of tomorrow. How do they propose to prove such a thing?

Current work centers on the logic gate, which can be regarded as the most elementary piece of computer hardware. In an electronic computer, the logic gate is a component that consists of any number of input wires and one output wire. The logic gate is a binary device: the signal in each wire is considered to be either a 1 or a 0. (In electronic terms, high voltage could correspond to a 1 and low voltage to a 0.)

Each logic gate is able to perform one of three basic operations: *not, and,* or *or.* The names of the three operations are based on the way the words *not, and,* and *or* are used in Boolean algebra, a pioneering system of formal logic developed in the 1840s by George Boole, the son of a poor cobbler. Self-taught in mathematics, Boole worked out a system of symbolic logic in which 1 stands for true and 0 for false. Although Boole's work earned him a mathematics professorship in Cork, Ireland, his system of logic was not fully appreciated by the mathematics community until some hundred years later, when the first electronic computer was built.

In formal logic (and ordinary English, too), the addition of the word *not* changes a true statement into a false one and vice versa. To put it in the terms of Boolean algebra, *not* converts a 1 into a 0 and a 0 into a 1. Thus, the *not* logic gate has a single input wire and converts the input signal into its opposite: if the input is a 1, it outputs a 0, and if the input is a 0, it outputs a 1.

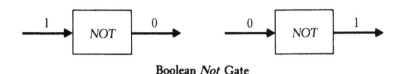

Boolean *Not* Gate

The word *and,* of course, serves to link individual statements into a composite statement, which is true if each of the components is true. To take a simple example, "Jules dined on Tofutti and Jim ate a Dove Bar" is a true statement only if both Jules and Jim dined on the aforementioned treats. By the same token, the *and* gate, which accepts two or more inputs, outputs a 1 only if all the inputs are 1's; otherwise, it outputs a 0.

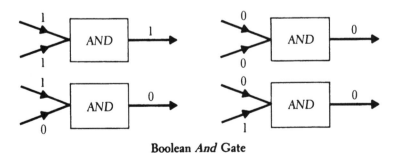

Boolean *And* Gate

The word *or* serves to link statements into a composite statement, which is true if one or more of the components is true. If either Jules or Jim (or both of them) ate their respective treats, then "Jules dined on Tofutti or Jim ate a Dove Bar" is a true statement. Similarly, the *or* gate, which has two or more inputs, outputs a 1 if at least one of the inputs is a 1.

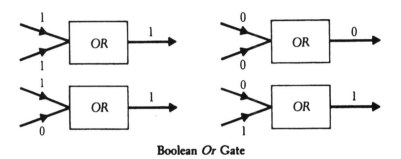

Boolean *Or* Gate

The beauty of Boolean algebra is that 1 and 0 need not stand for true and false but can represent any two distinct states. In the traveling-salesman problem, for example, 1 and 0 might represent the relevant relations between cities: a 1 if two cities are connected by a road, a 0 if they're not connected. In the clique problem, 1 could represent the state of two people's being friends (or, in the graphical representation of the problem, two points' being connected by a line) and 0 the state of their not being friends (two points' not being connected by a line).

In a computer, any number of *and, or,* and *not* gates can be linked together to form a circuit. The diagram below, for example, shows a small circuit of four *and* gates and one *or* gate that can solve a trivial instance of the clique problem: In a group of four people, are there three who are friends?

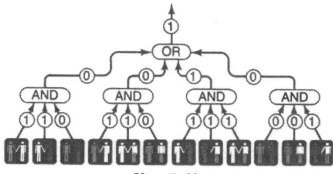

Clique Problem

As the size of the clique problem grows, however, the size of the circuit (that is, the number of logic gates) balloons exponentially for the *known* methods of solution. If mathematicians can prove that the circuit must be exponentially large for any conceivable solution method, known or unknown, they will have proved that the clique problem is immune to fast algorithms.

Not knowing how to start such a proof, mathematicians have looked instead at a particular problem—the parity function—that ordinarily has a fast algorithm and have tried to restrict the circuit in certain fundamental ways so that the fast algorithm no longer works. (The parity function determines whether there is an even or odd number of 1's in a string of 0's and 1's.) This approach may seem strange, but it isn't. Mathematicians know very little about how to prove that a circuit must be large, so any proof to that effect, even one in a contrived situation, would be a step forward and might provide the tools needed to prove the real thing. "This is a common approach in mathematics," says Sipser. "If a problem is big, try cutting it down some way and solving part of it, hoping that the partial solution will give insight into the original problem."

Early work in this area limited the depth of the circuits, where the depth is the number of levels of logic gates. The first results came in 1981, when Sipser and two coworkers at Carnegie-Mellon University proved that if they limit the depth of the circuit for the parity function, the width of the circuit blows up faster than any polynomial. In the spring of 1985, Andrew Yao at Stanford University sharpened this result by showing that the circuit's width blows up not only superpolynomially but exponentially—the desired sign that this problem, although artificially restricted, is intrinsically hard.

Word of Yao's result spread quickly through the mathematics community. "Everyone thought it was beautiful," says Sipser, "but also very complicated." With Yao's techniques paving the way, several researchers quickly improved upon his result. "It's like the first four-minute mile," says Ronald Graham, director of mathematical sciences at AT & T Bell Laboratories. "Once somebody does it, everybody does it."

In August 1985, Johan Hastad, a graduate student in computer science at MIT, took Yao's essential ideas but simplified the argument. "In the process," says Hastad, "I got a stronger result. The smallest circuits we know how to design [in this restricted problem] aren't much bigger than the size I have shown theoretically that they have to be." Each successive proof comes closer to showing that what mathematicians actually know how to write down for a circuit isn't much worse than the best they could do theoretically. For these restricted problems, it is the problem itself, not mathematical ignorance, that rules out a fast solution.

Two mathematicians at Moscow University, A. Razborov and A. Andreev, had much success in limiting not the depth of the circuit but the operations it can perform. Razborov proved that, if *not* gates are disallowed, the size of the circuit for the clique problem grows faster than any polynomial. And mathematicians here have improved this result to show that the circuit must blow up exponentially. By barring *not* gates, Andreev was able to show that another kind of problem also requires huge circuits.

These results have contributed to general optimism in the field, although no one yet knows how to go about dropping the restrictions on the circuits and proving that, unrestricted, the traveling-salesman problem is as hard as it seems. "There's still a long way to go," says Sipser. "Six years ago, I made a bet with someone—I hope he remembers—that the proof would be found by the year 2000. I'm still pretty confident—that's twelve more years." Graham is even more sanguine: "A proof in the next three years would not surprise me."

Despite the general optimism, researchers in complexity theory—the branch of mathematics that characterizes the difficulty of problems—have been known to have their intuition fail them. During the winter of 1985, David Barrington, a mathematics graduate student at MIT, proved that a certain primitive representation of what a computer can do is more powerful than anyone in the field had imagined. This primitive representation involves not *and, or,* and *not* gates but a branching gate that has two output wires. When a branching gate is triggered, if the input has a certain designated value, the gate sends a signal out along one of the two wires; for all other inputs, it sends a signal out along the other wire. In other words, a branching gate is able to handle such statements in computer programs as "If $x = 5$, go to step four; for all other x's, go to step seven."

Barrington proved that a circuit constructed entirely of branching gates with no more than five levels of gates can solve what is called the majority problem: in a string of o's and 1's, is there a majority of 1's? Complexity theorists had universally (and wrongly) believed that branching gates restricted to *any* fixed height—let alone the tough restriction of five stories—could not solve the majority problem.

"My proof is simple," says Barrington, "but it surprised everyone because they thought that what I was trying to prove was false." Barrington's result may have few practical applications—"except," he adds, "it may get me a teaching position at a good university." And it may persuade mathematicians not to be so cocky about their convictions in the complex field of complexity theory.

10

THE MACHINE WHO WOULD BE KING

We have looked so far largely at theoretical issues in computer science, at what kinds of computational tasks machines and men can do in principle. The kinds of limits we have discussed are absolute. If complexity theorists can prove what they suspect is true, the traveling-salesman problem can never be efficiently solved. It's not a question of mathematicians' and computers' lacking the proper tools; there are no tools to be had, and there never will be.

Most mathematicians and computer scientists are not up against theoretically insurmountable limits. The obstacles they face are self-imposed and can be circumvented, at least in principle. A major obstacle—and this looms large in many endeavors besides mathematics—is the tendency to play it safe by following the widely accepted, if not entirely successful, problem-solving techniques of one's colleagues. Those who strike out on their own had better succeed soon if they want to avoid the ridicule of their peers. In this chapter, we follow the pioneering efforts of Hans Berliner to build a computer that plays good chess. In the next chapter, we'll look at W.

Daniel Hillis's attempt to replace the basic architecture that has served the electronic computer well for its forty-year history with a revolutionary design of his own.

In his own, quiet way, Hans Berliner, a researcher in computer science at Carnegie-Mellon University, in Pittsburgh, wants to be the best in the world. He has had that distinction for himself, and now he wants it for his computer offspring. In 1968, he became postal chess champion of the world, after brilliantly demolishing the wily Soviet tactician J. Estrin in a forty-two-move game that had Berliner huddled over the board for five hundred hours. In 1979, a computer program he designed called BKG 9.8 beat the world backgammon champion, Luigi Villa of Italy, by the lopsided score of 7–1 in a highly publicized match in Monte Carlo. Like a proud father, Berliner was happy that BKG 9.8 had become the first machine to defeat a human world champion at any board or card game.

Today, BKG 9.8 is in mothballs (the world backgammon association has barred computers from official tournaments), but a rcokie program called Hitech, designed by Berliner and his graduate student Carl Ebeling, is upholding the honor of machines in another arena, that of the chessboard. In October 1985, Hitech won the North American computer chess championship. That success, along with a string of victories over talented humans, demonstrated that Hitech plays chess better than any other machine and better than 99 percent of the thirty thousand Homo sapiens ("thinking men") who play in tournaments sanctioned by the United States Chess Federation.

Berliner now has his eye on the Fredkin Prize—$100,000 to the designer of the first computer that defeats the human world champion. Hitech is currently not good enough to depose the champ. But, given Berliner's tenacity, tutelage, and track record, the program's future chances should not be underestimated.

Chronologically, Berliner's first love was chess; his second, machines. Born in Germany in 1929, he moved to America when he was eight and settled with his parents in Washington, D.C. Finding school to be much less demanding here than in Germany, he looked for challenges outside the classroom. In 1942, at summer camp, he saw some youngsters playing chess and asked them to show him the rules. "Even that first day," Berliner recalls, "I managed to find somebody whom I could beat. That was it. I was hooked."

Two years later, he was champion of his neighborhood chess club and holding his own at the best club in Washington. "My parents were not encouraging," says Berliner. "They warned me that I'd come to a bad end if I kept playing chess all the time. Who knows what would have become of me if no one had told me that?" In the short run, however, Berliner did not curb his chess playing. In 1949, he won the coveted Washington city championship at the tender, record-breaking age of twenty.

That same year, the American mathematician Claude Shannon wrote an influential paper in which he described in general terms how to program a computer to play chess. The electronic computer was only in its infancy then, but chess playing was already recognized as an important goal in the nascent field of artificial intelligence. Chess, unlike certain other intellectual endeavors, was appealing because the machine's competence at it could be precisely judged by pitting it against human opponents under controlled conditions. Tournament players have numerical ratings based on how they fare against other rated opponents. The computer would also earn a rating, reflecting its performance against rated human beings.

As pioneers in computer science tried to put Shannon's ideas into practice, the young Berliner concentrated on his own chess playing. By 1954, he was among the top twelve players in the country, and he stayed there for a dozen years. Sometime in the early 1950s, he read about the first efforts at computer chess. "Their games," he recalls, "looked pretty laughable to me."

One of the pioneers was the British mathematical phenom Alan Mathison Turing, a seminal thinker in artificial intelligence and (as we saw in chapter 8) a deep prober of the limits of mathematical knowledge. Like Einstein, Turing was an avid, if not accomplished, chess player; perhaps his lifelong fascination with the game stemmed from its being one of the few intellectual activities that he was unable to master. In any event, Turing wrote down half a dozen pages of recipe-like steps—in effect, a computer program—for mechanically playing chess. Although he never got around to encoding the chess-playing recipe into a computer, he used it to play a game in 1952 against Alick Glennie, a student at the University of Manchester in England who was a brilliant programmer and a less-than-brilliant wood pusher. Turing's paper machine (so called because it existed only on paper) lost that game, the first ever played by any idealized or realized machine.

Turing's recipe gave each piece a numerical value that more or less reflected its relative strength, as rated by chess textbooks: king 1,000, queen 10, rook 5, bishop 3.5, knight 3, and pawn 1. In choosing a move, all continuations that involved captures were followed until quiescent positions were reached, positions in which neither side could take a piece or deliver mate. For each of these quiescent positions, the relative strength of the armies was computed by adding up the values of the pieces, treating the values of the machine's pieces as positive numbers and those of the opponent's pieces as negative numbers. A move was selected that led to the quiescent position in which the machine maximized its relative strength.

Turing's evaluation scheme was able to find moves that led to the win of material but was of no use in static situations. For example, it didn't distinguish among first moves for White, because at the start of the game, none of the twenty possible moves (sixteen pawn pushes and four knight moves) captures a piece or even threatens to capture a piece, and so the twenty quiescent positions have the same relative-strength value of 0, which is clearly absurd.

Turing overcame this problem by also weighting, in static positions, such factors as mobility and king safety. A pawn, for example, gains 0.2 for each rank it's advanced beyond its home position, plus an additional 0.3 if it's defended by a piece other than a fellow foot soldier and minus 0.3 if it's undefended. The rook, bishop, knight, and queen each gains the square root of the number of legal moves it can make, plus an extra point if at least one of these moves is a capture. Moreover, should the rook, bishop, or knight (but not the queen) be defended, there's a bonus point if it's defended once and two bonus points if it's defended two times or more. The king gains 0.3 if it's castled, 0.2 if it's poised to castle, and 0.1 if castling in the future isn't prohibited.

Turing was also concerned with king safety. In his evaluation scheme, the king loses points depending on how wide open it is to attack. Turing ingeniously measured the wide-openness by imagining the king to be another queen and computing the mobility of this new queen. Moreover, Turing added 0.5 for a move that put the enemy king in check and a full point for a move that threatened immediate checkmate.

In static situations, the paper machine would make the move

that, according to the evaluation function, maximized mobility, the safety of its own king, and the vulnerability of the enemy king. In the 1952 game against Glennie, the paper machine opened with P–K4, the advance of the king's pawn two squares; of the twenty moves possible, P–K4 has the highest value (the move not only advances a pawn to the fourth rank but also enhances the mobility of the queen, the king's bishop, and the king's knight). As early as the third move, the paper machine made an inferior pawn sortie, but Glennie did not exploit it. On the twenty-ninth move, the machine greedily grabbed a pawn with its queen, because its evaluation function showed that Glennie had no immediate effective capture in return.

Paper Machine

The program overlooked a simple yet crushing rook move that, by pinning the program's queen against the king, led to the queen's forced capture. A proponent of cybernetic euthanasia, Turing resigned on behalf of the paper machine.

As primitive as the paper machine was, it did manage to do some things right. For example, it recognized that material considerations are of importance only after all captures are exhausted. In a position on the board you may be short a queen—which is normally an insurmountable disadvantage—but the chances are even if it's your move and you can capture the enemy queen. You would not want an evaluation procedure that merely tallied up the relative strength of the armies without taking into account the possible capture.

Turing was also on the right track when he incorporated into the evaluation function such aspects of chess knowledge as mobility and king safety. The machine's downfall in the game against Glennie was that it didn't have enough knowledge. It wasn't able to recognize the dangers inherent in a particular pattern of pieces: king and queen on the same file.

Berliner and other masters, and even numerous players of far less skill, have this pattern, and countless others, filed away in their minds. Studies show that the human master has a remarkable memory for chess patterns and positions but that this excellent memory doesn't necessarily carry over to matters unrelated to chess. On a personal level, Berliner found that the success he enjoyed on the chessboard didn't carry over into the classroom—at least not initially.

"I was one of those people who ran afoul in my early academic years," recalls Berliner. "I started off as an honor student in physics, but somehow I got completely turned off. I was working my way through school, and at one point I had saved enough money to pay for the rest of my education, so I stopped working. That was a critical error. All of a sudden I had a lot of time on my hands, so I took up bridge in addition to chess. I quickly became one of the top fifteen bridge players in Washington. Everything was going to pot."

After a stint in the army, Berliner wanted to go back to school. "I couldn't finish in physics," he recalls, "because my grade-point average was too low, so I switched to psychology. It seemed like a great field to study because there were all these interesting facts." Coming from physics, Berliner expected the facts to be organized into theories but was disappointed to find that this was not exactly so.

In 1954, he got married and, between home life and a new job, had little time for bridge but managed to keep up his chess. "I was working at the Naval Research Lab in a field called human engineering," he says. "It was pretty awful stuff, a mixture of psychology and physics, that dealt with the design of equipment. That was 1955, when computers were coming out, and the lab was building one. I had taken a course in programming, perhaps written a twenty-line program to add numbers, but other than that I had no contact with computers.

"Having no time to travel to chess tournaments, I decided to take up correspondence chess. That was another big mistake. It takes infinitely more time than over-the-board chess. For the next thirteen years, I played in postal chess tournaments and won them all. In the world championship, I had to play sixteen games. I figure I spent an average of four hours contemplating each move—and each game had some thirty-five moves. That means I invested 2,200 hours to win the title. Then I essentially gave up the game." He didn't relish the thought of spending another 2,200 hours defending his title.

In 1961, he joined IBM in Bethesda, Maryland, as a systems analyst and worked chiefly on contract work for the military. Although he spent eight years there and worked his way up to a position as a manager, he found the job unrewarding: "If you do it conscientiously, it's a terrible life. As a manager, you're the one who gets it from both the bottom and the top. You have people working for you who don't really give a damn, and you have a guy from the military telling you what to do who either doesn't really know what he wants or else wants something unreasonable. Then that guy gets replaced by another fellow, who doesn't understand what the first guy wanted and orders all sorts of changes. I started feeling that I wanted to do work that I could be proud of when I looked back on it. I wanted to go into research."

Berliner had been following computer chess from a distance, but the progress he saw was disappointingly slow. During the 1950s, academics made rosy predictions that belied the lack of success in the laboratory; in 1957, for example, Herbert Simon, now a Nobel laureate at Carnegie-Mellon, claimed that within ten years a digital computer would be the world's chess champion.

The magnitude of the programming task was not fully appreciated. According to popular wisdom, the chess master is a kind of human computing machine: when he chooses a move, he explores hundreds of continuations in his mind's eye—if I push the king pawn, he'll fork my rooks, but then I'll trap his queen—at lightning speed with incredible precision. Calculation is the stock-in-trade of computers, so it seems they should be naturals at chess. The problem is that the popular wisdom is wrong; calculation is not the only, or even the main, secret to the chess master's success. He depends much more on pattern recognition than on the exploration of a mind-numbing aggregation of moves.

The Dutch psychologist Adrian de Groot found that of the thirty-eight legal moves possible in the typical position, the master ponders an average of only 1.76. In other words, he is generally choosing between two candidate moves that he recognizes, based on the hundreds of thousands of positions he has played himself or seen others play, as contributing to the immediate and long-term goals of the position. "Calculation," writes Father William Lombardy, a U.S. grandmaster, "most often comes after the goal is achieved, the moment when a winning position converts into a mathematically forced win." The instant recognition of familiar patterns is what enables the grandmaster to play a remarkably strong game of chess when given only a second or two a move; in this kinetic version of chess, there simply is no time to look ahead.

Many of the early programs limited themselves to pondering only a select number of candidate moves (although never anywhere near as few as 1.76). The problem with the selective-search approach is that no one knows how to express in computer language, let alone in English, general, fail-safe principles for choosing candidate moves. In 1966, the most successful of the early selective-search programs, MacHack, developed by Richard Greenblatt at MIT, became the first machine to defeat a human player (albeit a weak one) in tournament play. MacHack also had the pleasure of trouncing Hubert Dreyfus, the author of *What Computers Can't Do,* a man who has made an academic career out of belittling the ability of machines.

In general, however, MacHack's play was seriously flawed. Although it was capable of playing competent chess for a long stretch of moves, it was apt to suddenly make a catcalling blunder that was somehow sanctioned by the general principles of chess programmed into it. Moreover, it sometimes overlooked subtle but effective moves that defied those principles. But it had defeated human tournament players—and that was a milestone in computer chess.

"When I heard about MacHack's victories," says Berliner, "I thought, my God, after all this lethargy in computer chess, after all these well-placed people trying various things with little success, there's a glimmer of hope. I went to see Greenblatt, and, although I didn't understand enough about computers to really follow what he was doing, I was impressed. Because I was between marriages, I once again had a lot of time on my hands, so I taught myself programming and spent my evenings and weekends writing a chess program. I asked IBM if I could work on computer chess at their research facility in Yorktown Heights, New York. They said, 'That's not the kind of project we're funding. Besides, you don't even have a Ph.D., so we could at best let you do that on the side if you were doing something else that was useful to the company.'

"I decided that the only way I could make it was to join them by getting a Ph.D. I had an exaggerated opinion of my background. I applied to several schools, but Carnegie-Mellon was the only one that accepted me." His victory in the world correspondence chess championship in 1968 apparently helped him get in.

"So there I was in the fall of 1969, a student at the age of forty. It was a big shock. I found that I had an awful lot to learn, about the theory of automata, different programming languages, various hardware configurations, and artificial intelligence itself." But whereas in his earlier years in academe Berliner had disliked his courses, he now took to them.

At Carnegie-Mellon, Berliner continued to work on the program that he had started in his spare time at IBM. Called J. Biit (pronounced "jay-bit," an acronym for Just Because It Is There), the program turned in a respectable performance at the first U.S. computer chess championship, in New York City in 1970. Like MacHack, J. Biit did a selective search. The program's strength was its evaluation function—how it numerically weighed the strength and weakness of each position it looked at—but because it searched selectively, it sometimes didn't even consider the right move, let alone make it. "In specific instances," says Berliner, "it could play brilliantly. But that's not enough. You have to be consistently right, in all different kinds of positions. J. Biit just wasn't robust enough to confront successfully the whole game of chess."

In that first U.S. championship, J. Biit lost to the program Chess 3.0, designed by the Northwestern University graduate students David Slate and Lawrence Atkin. A subsequent version of Chess 3.0 conducted not a selective but a full-width search: the exhaustive analysis of all possible continuations to a certain prescribed depth. While a full-width search always includes the right move among the candidate moves it looks at (since it looks at everything!), it is a very inefficient way to choose a move. Much time is squandered on the exploration of horrendously bad continuations that even the weakest of human wood pushers would never ponder for an instant. The wasted effort would hardly matter if the computer could see clear to the end of the game, as it can in, say, ticktacktoe.

The mathematics of chess is testimony to the inefficiency of a full-width search. A game between human masters typically takes 84 plies (a ply being a move for a given side). Since there are an average of 38 legal moves in each position, an exhaustive search would have to consider 38^{84} possible positions. That's an extraordinary number of positions: 38^{84} is bigger than 10^{132}, the number 1 followed by 132 0's. The universe has been in existence for only on the order of 10^{18} seconds, and so a computer working for as long as the age of the universe would have to analyze 10^{114} chess positions per second in order to see clear to the end of the game.

In tournament play, computers, like people, are not allowed to think for eternity; they are given only about 120 minutes for forty moves, which amounts to 3 minutes a move on the average. Even if the machine adopts the much more modest goal of exploring all possible continuations to a depth of only a few moves, the mathematics is forbidding. After only two plies (a move for each side), the number of possible positions exceeds a thousand. After four plies, there are more than a million possible positions.

The computer must not only generate all these positions but also evaluate them. It does this rather crudely by numerically weighing such factors as material (the number and nature of the pieces and pawns on each side), mobility, control of central squares and open files, pawn structure, king safety, and so on. At the end of, say, three minutes, it makes whatever move will minimize its opponent's potential maximum gain; this mini-max approach, borrowed from the mathematical theory of games, assumes that the opponent sees everything you see and is out for his own self-interest.

A full-width search, then, even if limited to a depth of only a few plies, would be impractical, were it not for the discovery of the alpha-beta algorithm, a clever approach to evaluation that lets the computer choose its move without having to evaluate every possibility. Yet, amazingly, the chosen move is the same move the computer would have made had it looked at every continuation. How is this possible?

Suppose the machine first explores all the continuations of a specific candidate move, call it A, to a certain depth. Assuming best play for both sides, the computer assigns a minimax value to A of, say, 1. (In this scheme, positive values correspond to an advantage for the computer and negative values to a disadvantage; an advantage of 1 represents the value of having one more pawn than the opponent, all other things being equal.) Now, the computer starts to evaluate another candidate move, call it B, a particularly stupid move that puts the queen on a square where it can immediately be captured by a lowly pawn. If the computer now examines the opponent's natural reply, pawn takes queen, and rules out the slim chance that it has brilliantly sacrificed its queen for an unstoppable attack, it will assign a numerical value to the position, say −9, indicating that its opponent has a huge advantage.

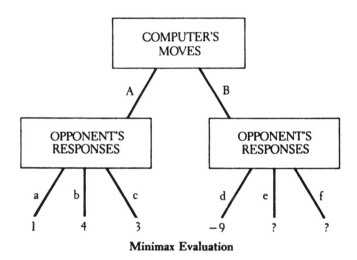

Minimax Evaluation

Modern computer chess depends on the minimax approach: make the move that minimizes your opponent's potential maximum gain. Suppose the computer has a choice of moves A and B. It sees that the opponent's best response to A is move a. (The numbers in the diagram show how good the resulting position is from the computer's point of view.) The computer now considers B and sees that the opponent, by responding d, can ensure himself a better result against B than against A. The computer now knows enough to choose A, no matter what the result of reponses e or f.

The computer does not need to consider the consequences of all other replies, in which the opponent doesn't take the queen, because it has identified a line of play in which the

opponent can ensure himself of a better result against move
B than against move *A*. Therefore, the machine knows that
from its own point of view move *A* is preferable to move *B*.

To implement the alpha-beta algorithm effectively, the computer must look at moves in an orderly fashion; in the above
example, it must examine *A* before *B*, and in analyzing *B*, it
must examine the capture of the queen before looking at other
responses. The order in which it investigates moves is dictated
by various heuristics, or general rules of thumb.

The capture heuristic, for example, instructs the program to
give high priority to moves that involve the capture of pieces.
(Such a capture, which has a better than average chance of
being a good move, particularly if the captured piece is undefended, also benefits the computer by helping it clear its
brain; with one fewer piece on the board, it has fewer replies
to consider.)

The killer heuristic keeps track of which of the opponent's
responses killed, or refuted, a particular move. When another
move is contemplated, the killer responses are investigated
first. Let's take an extreme case. The computer discovers that
its contemplated capture of an enemy rook is refuted by the
opponent's delivering checkmate. In pondering an alternative
move, it will determine first whether the move avoids the
checkmate. In other words, the killer heuristic serves to identify and monitor threats, here the lethal threat of an immediate
mate. Another heuristic assigns a high priority to moves that
deliver check, in accordance with the old aphorism "Always
check, it might be mate." In short, the computer now behaves
a little more like a human being.

More economy can be achieved in a full-width search by
looking progressively deeper into all continuations rather than
diving fully into them one at a time. From the position on the
chessboard, all possible continuations are examined first to a

depth of one ply, and, on the basis of the search so far, the best move is noted. Starting with that move, all continuations are then effectively examined to a depth of two ply, and a best move is again noted. The process, called iterative deepening, is repeated until the desired depth is reached.

The effectiveness of a full-width search can also be enhanced by a table that keeps track of positions the computer has already evaluated, the values assigned to them, and the best move found so far. In a full-width search, positions tend to come up more than once, and the table is a useful time-saver, provided it is designed so that it takes the program less time to look up the evaluation than to recompute it.

In the 1970s, Slate and Atkin at Northwestern were able to make subsequent versions of Chess 3.0 work successfully with, among other things, minimax evaluation, the alpha-beta algorithm, capture and killer heuristics, iterative deepening, tables of positions already examined, and—as in Turing's paper machine—a deepening search of tactical lines of play until a quiescent position is reached. The result was a program, Chess 4.7, that played competent chess slightly below the master level.

In 1981, the full-width-search program Belle, created by Ken Thompson and Joe Condon of AT & T Bell Laboratories, became the first computer to achieve a master rating, which put it in the top 1 percent of all U.S. tournament players. Belle owed its success to custom-built hardware designed specifically to execute chess calculations. Officials in Washington apparently thought very highly of Belle. In 1981, federal agents apprehended Thompson and Condon when they tried to take Belle on a plane to Moscow to play an exhibition match. The Reagan administration feared that the program might give away military secrets. Thompson insisted that the only thing

Belle knew how to do was play chess. "The only way Belle could be used militarily," Thompson told the press, "would be to drop it out of an airplane. You might kill someone that way." These days, Washington is less impressed because Belle's rating has slipped below the master level, but it still plays formidable chess by looking an average of eight plies deep, analyzing 120,000 positions per second.

While Slate, Atkin, Thompson, Condon, and others were getting the full-width search to work, Berliner was concentrating on the evaluation function. "I was thinking," he recalls, "about de Groot's well-known studies of how masters play chess—how they look partway into a variation, then turn to something else, then go back to that first variation. It seemed right. At least that's how I thought I played chess." The existing computer programs, on the other hand, did not move back and forth between variations. They followed a particular variation to a certain depth, assigned a numerical value to the resulting position, and moved on to another variation.

"The trouble with assigning a specific value is that you can't afford to be wrong," Berliner says. "Take a position where you've sacrificed two pawns to obtain a very strong attack. If you use something like alpha-beta, you come to a terminal position to which you have to assign a value; either it was worth giving up the pawns for the attack or it wasn't. Whichever view you take, you're going to be wrong a certain amount of the time. It would be much better to say, 'I'm still not sure. I've lost two pawns and have a very strong attack. I might in fact mate the guy or win back more than the two pawns, but I might also simply be out two pawns.' So you leave the question open and probe a little bit deeper to see if you can resolve it.

"I was thinking a lot about this sort of thing, about how to make the program look deeper when it should. One evening,

I had an inspiration: Instead of assigning a single value to a position, why not a range of values?"

The top value in the range would mean that the position was at most that good, and the bottom value would correspond to the worst it could turn out to be. The program would compare ranges of values rather than single values, and when a range was too wide, it would look deeper into the position in order to make the top and bottom values converge. "This idea was the missing ingredient," says Berliner. "It's one of those miraculous things that happens in science every once in a while. You come up with one thing, and, suddenly, everything makes sense." The idea of using a range of values became known as the B* (pronounced "B-star") algorithm, and Berliner filed it away in his bag of tricks.

In 1975, as Berliner finished his doctoral thesis on computer chess, he decided to program a machine to play backgammon, a game he had recently learned from his new wife's father. He found the domain of backgammon an attractive one for studying evaluation because it is a game in which searching won't get you very far. In a typical backgammon position, there are more than 400 possibilities (twenty-one dice rolls and some twenty ways of playing each roll), compared with "only" 38 possibilities in the typical chess position.

In the backgammon program BKG, Berliner departed from the standard practice in artificial intelligence of doing evaluation according to rules. "In a medical diagnosis system," notes Berliner, "you might have a rule that says, 'If the patient has such-and-such disease and is over six years old, then you should give thus-and-such treatment.' And then all of a sudden you're faced with a patient with the disease who's five years nine months old. According to the rule, you can't give the treatment. Well, that's wrong. What you really want is not a black-

and-white cutoff point but a smooth function that somehow takes into account such factors as age, weight, and general health. In this particular case, you might want to prescribe a reduced dosage.

"When you're first designing intelligent systems, these kinds of considerations recede into the background. They're not nearly as important as just getting the basic information into the machine. But if you want to compete with the best humans, you can't operate by a set of rules that's totally inflexible."

Berliner, of course, wanted his backgammon program to hold its own against the best human players, and so, after flirting with more conventional approaches, he ended up spurning rules that would divide positions into classes, each class having a different evaluation function. Instead, he relied on a single mathematically complicated function that includes some fifty different variables, corresponding to particular features that have different degrees of importance, depending on the stage of the game. Each variable is replaced by a number, a measure of the extent to which the corresponding feature is present in the given position. Then each number is weighted: it is multiplied by another number, called a coefficient, which represents how much (or how little) attention should be given to that feature at that point. The coefficients change slowly and smoothly as the game progresses.

The success of this approach, called SNAC (for smooth, nonlinear function with application coefficients), was evident when BKG trounced the human world backgammon champion only a few months after SNAC was introduced. Although the program had some lucky dice rolls, and got away with a few small errors, it showed itself to be a great player.

From his success at computer backgammon, Berliner knew that having a smoothly varying function would also be the key

BKG's Winning Plays against Luigi Villa, the Human World Backgammon Champion

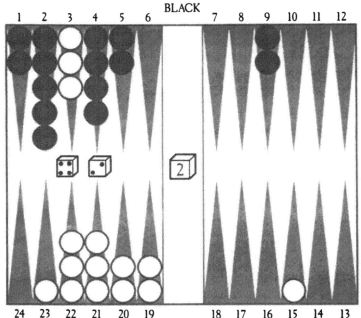

BKG ROLLED A FOUR AND A TWO *in this position from the first game of its match with Villa. The program* (Black) *had the advantage but was forced to leave a blot. It played 9-5 and 9-7, leaving a blot on the 7 point that could be hit by 13 dice rolls. The apparently safer play of 5-1 and 4-2, leaving a blot on the 5 point that could be hit only by 11 rolls, is inferior because it leaves two pieces on the 9 point that could become exposed later when they must move.*

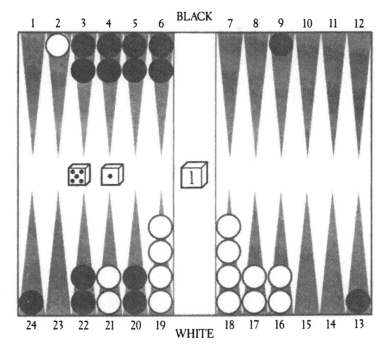

BKG ROLLED A FIVE AND A ONE in this position in the last game. The program made the sensational play of 13-8 and 3-2. If any of BKG's blots were hit, it would have more time to build up its back game. On the other hand, if they were not hit, it would be able to make points in its home board, making it more difficult for Villa's piece to come back in and then escape.

to effective evaluation at chess. Here, too, the conventional approach involved rules. Consider king position. In the middle game, you want your king tucked away in the corner, where it's less likely to be harassed. The evaluation function might count the number of squares between the king's actual position and the corner; the greater the number, the worse off you are. In the endgame, however, when there are so few pieces left that the danger of being mated is slim, the king should be in the center of the board, where it can function as a strong fighting piece. So in the endgame the evaluation function might count the number of spaces between the king's actual position and the center. If you use a rule like "When there's a certain number of pieces and pawns on the board, it's a middle game and when there's one unit less, it's an endgame," you get schizophrenic behavior at the boundary.

"You don't want this," says Berliner. "It should be continuous—the middle game turning gradually into the endgame. As the endgame approaches, you're no longer so certain that you want the king in the corner, and you tolerate the king migrating slowly toward the middle of the board. When everyone has agreed that the endgame is finally here, the king should be close to the center, not hiding off in the corner." The way to achieve this is to have one smoothly changing evaluation function, rather than an arbitrary distinction between the middle game and the endgame and a different evaluation function for each.

By May 1985, Berliner had incorporated many of his ideas about chess (although not the B* algorithm) into the program Hitech. The hardware for Hitech consists of a $25,000, off-the-shelf Sun computer and a tray of sixty-four identical special-purpose chips, designed by the graduate student Carl Ebeling. The chips were manufactured by VLSI (very large-scale integration), a state-of-the-art technology for cramming 15,000 circuit elements onto a single chip. The more elements, the faster a chip can generate and evaluate chess positions.

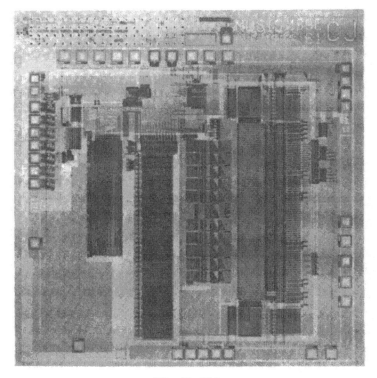

Hitech's Chip

Each chip is responsible for one square of the chessboard, monitoring what pieces or pawns can move there and doing a preliminary ordering of those moves. All the chips operate independently and simultaneously, which saves even more valuable time. The evaluation of positions is done by eight additional parallel hardware units, each capable of evaluating some aspect of the whole position. These units depend on information downloaded from the Oracle, a program running on the Sun that is the source of Hitech's chess knowledge.

The secret of Hitech's success is that it thinks better (because of the Oracle), as well as 50 percent faster (because it simultaneously evaluates more than one move sequence), than its mechanical rivals. Conducting a full-width search, Hitech examines on the average an astounding 175,000 positions per second, or 30 million in the three minutes allotted for each move. "I doubt," says Berliner, "that human beings look at 30 million alternatives in the course of their lives."

Hitech's speed and brains make it the highest-rated chess program in the world and better than all but a Kremlin full of humans. Berliner thinks that Hitech or one of its descendants has a fifty-fifty chance of deposing the human king of chess by 1990. In pursuit of that goal, he plans to put more knowledge into the Oracle and have Hitech try a selective search, perhaps even guided by the B* algorithm.

How good will Hitech get at chess? For that matter, how good will any machine ever be at a given intellectual activity? "I think," says Berliner, "that we're going to discover that there's a limit to how much knowledge you can stuff into a machine before some bit of that knowledge starts contradicting another bit." Some researchers are trying to defuse this possibility by introducing a belief system—the machine pays less attention to one piece of contradictory knowledge because it comes from a less reliable source.

"But I don't think a belief system is the answer," says Berliner. "I think that we want to build a learning machine, something that's going to sit on the shelf, watch video tapes, and learn from the bottom up. It may learn very slowly at first. It may take twenty years for it to have the understanding that an adult human being has. And that'll be fine. If what it has is worthwhile, it'll be worth it. But I'm not holding my breath. It is bound, however, to happen eventually—not this decade, perhaps, but the one after."

11

A BOY AND HIS BRAIN MACHINE

It is early morning in Cambridge, Massachusetts, in the summer of 1986, and W. Daniel Hillis, the thirty-year-old founding scientist of the Thinking Machines Corporation, is slumped in a chair, staring wearily at a blank video display screen. He enters a few commands on a keyboard and an image of black and white lines resembling a dartboard appear on the screen. As Hillis punches the keys, a sleek, black, five-foot, glasslike cube in a room down the hall—a computer designed by Hillis and known as the Connection Machine—is punctuated by thousands of tiny red lights that flash frantically in no discernible pattern.

In this apparent randomness, however, may lie the future of computing. "Last night," he says, "we had a breakthrough. The machine actually learned. It learned by itself, without my ever telling it whether it was right or wrong."

Hillis and a colleague have spent a long night programming the Connection Machine to unscramble slightly distorted black-and-white line images that have been fed into it. The process is a primitive example of something called visual adaptation, which humans do very well. "If I slipped a crazy pair of glasses on you that distorted your vision," says Hillis, "you would learn to see normally." But most computers, unlike people, do not learn from experience.

That night the Connection Machine was an exception. After receiving a distorted image, the computer would display what it thought the real image looked like. Hillis never told it how well it was doing. Unlike a chess-playing computer, which does not get better from game to game unless the programmer tinkers with it, the Connection Machine had improved each time. After a couple of hundred tries, it was getting the image pretty much right. After three minutes, or five hundred tries, it was totally undoing the distortion.

For Hillis the breakthrough was not that the Connection Machine could do visual adaptation—although such a skill might be useful in interpreting fuzzy photos or even, conceivably, in cryptography, where scrambled messages must be unscrambled—but that it had *learned* to do this. If it could learn to do this, it could undoubtedly learn to do other things, too. And that, Hillis thinks, is essential if true artificial intelligence is ever going to be more than a dream.

The Connection Machine is the most dramatic example of an emerging breed of computer, the parallel processor, that is beginning to transform computer science. Traditional computers, even powerful ones, rely on just a single processor, the computational engine where the calculating takes place. The Connection Machine is radically different; it harnesses the collective power of 65,536 small processors, or minibrains, all working in concert to solve a problem.

A parallel processor is simply a computer that has more than one processor. The underlying principle is rather simple: two heads are better than one. And if two heads are better, why not 4 heads or 16 or even 65,536? In theory, the additional heads, or processors, speed up the computer's performance, enabling it to tackle not only problems in artificial intelligence, involving vision and speech understanding, but also a host of numerically intense problems faced every day by physicists, engineers, and military planners.

In a way, Berliner's chess-playing computer is a parallel processor; it has sixty-four chips, each of which corresponds to one square on the chessboard. However, these chips can evaluate only chess moves, whereas Hillis's processors are flexible enough to tackle all sorts of computational problems.

As simple as the concept of parallel processing may sound, formidable obstacles stand in the way of turning the idea into silicon. How many processors are optimal? How smart should each individual processor be? How should the processors be connected so that they communicate efficiently and work in concert?

There is also the difficulty of how to program, or instruct, the processors to solve a particular problem. Some problems are like Tom Sawyer's task of painting a picket fence; it's easy to see how the job could be divided among additional laborers. Other tasks are more akin to Mark Twain's writing *Huckleberry Finn;* it's not evident how Twain could have benefited from the assistance of other writers.

The Connection Machine is a working example of one way to confront these difficulties. In August 1986, Thinking Machines delivered a scaled-down, $1 million version of the Connection Machine, with 16,384 processors, to its first commercial customer, Perkin-Elmer. The machine was installed at MRJ Inc., a Perkin-Elmer think tank in Oakton, Virginia, which does contract work for NASA and the Department of Defense. "After using the machine a few weeks," says the MRJ staff member Tom Kraay, "we solved an important military problem": Given that you know the position of enemy radar and a destination you want to reach, what flight path should you take to minimize the chance of being detected? "This problem comes up a lot," says Kraay, "like when we bombed Libya." Common as it is, the analysis is numerically intense and a general solution has proved elusive.

Hillis's company, the Thinking Machines Corporation, was founded in May 1983. At the time, companies involved in artificial intelligence—machines that do things that people are inclined to call intelligent—were working on expert systems, computers that can mimic human experts at one particular activity, say, deciding what chess piece to move, what bonds to buy, or where to prospect for oil. Expert systems are still the rage, and despite all the hoopla in the media and on Wall Street about artificial intelligence, the best expert systems are nothing more than idiot savants; a chess-playing computer, for example, can do nothing other than play chess.

Thinking Machines was formed with the long-range goal of building not an expert system but what Hillis calls an amateur system: a machine that has common sense. As the company's glossy promotional brochure puts it, "Someday we will build a thinking machine. It will be a truly intelligent machine. One that can hear and speak. A machine that will be proud of us." If this hyperbole had been the only thing the company had going for it, it might not have gotten off the ground. But as a means to its quixotic end, Thinking Machines had the short-term goal of building the first massively parallel processor.

Even this goal was ambitious, but Hillis at least had an idea of how to carry it out. Moreover, even those who were skeptical about artificial intelligence were intrigued by the possibilities of parallel processing. Thinking Machine's president, Sheryl Handler, a woman in her thirties who had helped launch Genetics Institute Inc., a pioneering biotechnology firm, decided that the best way to achieve these goals was to bring together an advisory cast of scientific superstars. Today, the company's advisers include the MIT professor Marvin Minsky, one of the pioneers of artificial intelligence; the Nobel Prize–winning physicist Richard Feynman, who served on the presidential commission that investigated the *Challenger* disaster; Jerome Wiesner, former president of MIT and science adviser to John F. Kennedy and Lyndon Johnson; and Stephen Wolfram, a young physicist who was at the Institute for Advanced Study and who had published his first scientific paper at the age of fifteen. Even the workers in the company's gourmet lunchroom, where Hillis and the others dine on carrot vichyssoise, bagna cauda salad, plum clafouti, and pavlova, are brainy; one kitchen worker left the company because he won a Fulbright grant.

With its all-star lineup of headstrong academics, many outsiders saw Thinking Machines as a highbrow think tank, full of romantic ideas about artificial intelligence but short on the nuts-and-bolts know-how needed to build an eggbeater, let alone a novel computer. Yet, with $16 million from William Paley, the founder of CBS, and other investors and with $4.7 million from the Department of Defense's Advanced Research Projects Agency, known as DARPA, Thinking Machines managed to design and build the Connection Machine in only two and a half years.

A puckish six-footer with twinkling eyes, Hillis does not look like the sort who would create a revolutionary computer architecture. His office, a few blocks from his alma mater, MIT, looks more like a playpen than a high-tech workplace. Near his captain's desk is a pile of Japanese mechanical toys, a cow-size cardboard dinosaur, and a propeller-driven wet suit that he made so that he could walk on water. The wet suit may not even be the craziest thing he has built; as an undergraduate, he constructed a huge ticktacktoe-playing machine from fishing tackle and thousands of Tinkertoys. Toys and gadgets, he claims, help him relax and clear his head.

"I want to build computers," Hillis says, "that are based, even loosely, on the structure of the human brain. The brain doesn't have one processor, like a conventional computer. It has a great many things—neurons—working in parallel. That's why the Connection Machine was designed to be massively parallel."

Hillis is by no means the only one giving computers additional processors. Perhaps a hundred other projects to build multiheaded computers are under way at universities and companies, chiefly small start-ups funded by venture capital. Many companies bill themselves as selling parallel processors, but industry analysts disagree about who is offering the real thing— the processors must be able to attack one task together rather than work independently on different tasks. (The domestic equivalent of truly parallel processing is mom and dad preparing dinner together; it's not enough, while mom is cooking dinner, for dad to be balancing the checkbook, however useful that may be.) IBM, the world's largest computer company, is also spending tens of millions of dollars on research in this area and hopes to have two experimental designs up and running in 1987. Hillis, however, has succeeded in linking up tens of thousands more processors than anyone else.

Despite the flurry of activity in parallel processing, the technology is still in its infancy. Nevertheless, there's wide agreement in computer science that parallel processing will be the technology of the future. In 1980, Japan announced its Fifth Generation project, a $1 billion, ten-year national commitment to building a new kind of computer that can easily converse with people and interact with its environment; parallel processing, the Japanese said, was central to this effort. American government and industry reacted with alarm. DARPA, formed in the wake of Sputnik to make sure the United States never again fell behind in frontier technology, announced the Strategic Computing program, its own version of the Fifth Generation project. As part of this initiative, DARPA planned to spend $70 million.

The soul of the old machine—the conventional, single-processor architecture—was John von Neumann, a brilliant mathematician who did seminal work in quantum mechanics, ballistics, meteorology, game theory, and nuclear-weapons design. When he came up with the single-processor architecture, in the 1940s, he wasn't being lazy or myopic in his vision of computing. The technology simply didn't exist for him seriously to propose building more than a single processor. Since neither the transistor nor the microchip had been invented, the earliest computers were built from ungainly vacuum tubes; even with only one processor, the 1946 ENIAC, the first general-purpose computer, filled an entire room.

In von Neumann's design, the processor was separate from the computer's memory, which contained not only the data for a particular problem but also instructions for manipulating that data. The separation made sense in the 1940s because two different technologies were involved. The processor was made of comparatively fast and expensive vacuum tubes, whereas the memory was made of much slower, inexpensive mercury delay lines. Von Neumann's idea was to program the computer so that the speedy vacuum tubes were busy and the slowpoke memory was relatively idle. This required the programmer to break up a problem in such a way that it could be solved step by step—as Hillis puts it—by "streaming memory past the processor." Chunks of data and instructions were shunted back and forth between the processor and the memory along one narrow pathway.

Today, the sharp division between the processor and the memory no longer makes sense, even though it's still found in all but a few pioneering machines. The processor and the memory are now both made of the same material—silicon. Despite the change in technology, the idea of keeping the processor busy by having it attack a problem step by step hasn't changed at all in conventional machines. The result is a tremendous inefficiency: 97 percent of the silicon, which is the part devoted to memory, is generally idle, while the mere 2 or 3 percent devoted to processing is frenetically busy. Hillis decided to find a way of better utilizing the memory and an alternative to the one-step-at-a-time approach to problem solving.

What Hillis had that von Neumann didn't was a source of small, cheap processors. In 1970, a tiny start-up company in Santa Clara, California, called Intel (short for *In*tegrated *El*ectronics or, less modestly, *Intel*ligence) managed to cram the 2,300 components of a processor onto a single slab of silicon an eighth of an inch long by a sixth of an inch wide. The microprocessor, or "computer on a chip," was born. A computer that occupied an entire room in the 1940s and 1950s was now the size of a thumbnail.

Intel and other companies soon figured out how to mass-produce microprocessors, making computing power almost as widely available to industry as electricity or water was. In 1975, there were 750,000 microprocessors in existence. In 1985, there were 353 million. In 1990, there will be 1.2 billion, according to Dataquest, a market-research firm.

While Intel was carrying the torch for miniaturization in the early 1970s, another fledgling company, Minnesota-based Cray Research, was moving in the opposite direction. Seymour Cray, the company's reclusive founder, set out to build the world's fastest computer by combining chips to create one mammoth, ultrapowerful processor. The processor in his first supercomputer, the Cray 1, is shaped like a huge letter C, six feet tall and nine feet in diameter at the widest. Five to ten times as fast as any machine then in existence, the Cray 1 gave off so much heat that it would have burned through the floor had the resourceful Cray not thought to snake freon tubes through it; old-fashioned refrigerator technology saved the day.

Cray Research has built two-thirds of the 180 supercomputers now in existence. The Cray 2, which has four processors and adopts some extremely limited elements of parallelism, is currently the world's fastest computer, six to twelve times speedier than the original Cray. Although these machines are appreciably faster than microprocessors, they are disproportionately expensive. The Cray 2 is five thousand times faster than a simple microprocessor, but at $20 million it is several hundred thousand times more expensive. This brute economic fact is one of the main reasons the government, universities, and many companies are pursuing parallel processing, even if the technology never contributes to artificial intelligence, as Hillis believes it will.

The stakes in parallel processing are high. Industry and government have taken for granted that ever more powerful computers will continue to be built year after year. In the past four decades, single-processor computers have been speeded up by a factor of a thousand, chiefly by shrinking the basic electronic components and packing more of them closer and closer together. Further dramatic speed-ups in single-processor computers may not be possible, however, because the design is up against fundamental physical limits, such as the reality that no signal in the circuitry can move faster than the speed of light. Significant increases in performance may come only from harnessing the power of more than one processor.

A single processor is simply too slow to do all the things a computer has to do to be intelligent. A truly intelligent computer—the "amateur system" Hillis speaks of—must be able to see, to understand human speech, to read English text, to reason, and to plan. "All these things are hard for a single-processor machine because they require a tremendous amount of information," says Hillis. "If you try to make the machine smarter by giving it more information, you actually make it dumber because it's much slower in accessing that information." A single-processor machine responsible for guiding an unmanned military vehicle would be worthless if it needed a year to "see" the difference between an enemy tank and a boulder. Parallel processing may be the way out. Speed can be maintained by dividing the information among different processors.

It is important to realize that, in a certain theoretical sense, the only advantage a parallel processor offers is speed. In 1937, Alan Turing—whose contributions to the theory of computing are second only to von Neumann's—demonstrated in effect that any computer, given sufficient time and memory, can do anything that any other computer can do. Therefore, any program that could conceivably run on a parallel processor, even if it had oodles of processors, could be mimicked on a single-processor machine, albeit ploddingly. In theory, then, all computers are equal.

In practice, however, scientists want computers that do things in real time. They want, for example, to be able to talk with a computer at a normal conversational pace, not having to wait eons for it to respond to a statement. The promise of doing things in real time is what parallel processors offer over single-processor machines—not only for problems in artificial intelligence but also for a host of thorny computational problems in climate modeling, fluid flow, plasma physics, subatomic particle physics, battlefield management, and the Strategic Defense Initiative, President Reagan's space-based missile defense plan, popularly known as Star Wars.

Danny Hillis was born in Baltimore, the son of a U.S. Air Force physician who moved around the world studying hepatitis epidemics. Hillis tagged along, building zany contraptions everywhere he went. He made a solid-fuel rocket to send grasshoppers aloft. He turned a tin can and a rotisserie motor into a mobile robot. Even at MIT, where Hillis did both his undergraduate and his graduate work, he continued to build wacky toys, such as a wand that spoke when you waved it in front of someone.

It was during his freshman year, in 1974, that Hillis began collaborating with Marvin Minsky, who was then forty-seven. Minsky has been active in artificial intelligence since the field began forty years ago. Like Hillis, Minsky is not only a deep thinker but an ingenious tinkerer. In 1951, he built one of the first electronic learning machines, from three hundred vacuum tubes, a bunch of motors, and a gyropilot from a B-24 bomber; like a rat in a psychology experiment, the machine learned to "run" a maze. In 1956, Minsky and three colleagues organized the first conference on artificial intelligence—at which the field was officially launched—and two years later he cofounded MIT's Artificial Intelligence Laboratory, dedicated to building machines that did such nonnumerical things as reasoning by analogy.

The young tinkerer and the old pioneer met for the first time when Hillis wandered unannounced into Minsky's office. The veteran computer scientist was trying to build an inexpensive computer that could do good graphics, and the machine's innards—its elaborate circuitry—were scattered on his desk. "This freshman appeared," Minsky recalls. "He was hanging around the office. We started talking about something or other, while he looked at the circuits. Then he said, pointing to one of them, 'I don't see why you need *that* 'cause the circuit over there does the same thing.' I looked, and, sure enough, he was right. I was suitably impressed because he was using just intuition—he didn't know what the circuits were for! It was clear that this freshman was something else."

The Connection Machine is the outgrowth of a question that has loomed large in Hillis's mind ever since the time of his first discussions with Minsky: Why can't a machine be more like a man?

"There are all sorts of things that people can easily do that machines can't," says Hillis. "You can make a machine that is very good at putting a tiny pin into a tiny hole with a precise amount of force. Yet, in spite of that precision, the machine can't pick up a glass of water without spilling it. That's sort of a paradox—the machine is much more precise than a person, and yet it's much clumsier." Those who thought about the paradox before Hillis did usually attributed the person's success to his having an accurate image of the environment he's interacting with. Hillis found this explanation superficial; a person's image of the water glass was not static, he thought, but constantly being adjusted because of tactile feedback: "If you look at how we pick up a glass of water without spilling it, it doesn't have anything to do with how precisely we position our hand or how precisely we apply a force. It has to do with our getting very good feedback from our fingers—we can do this even with our eyes closed, by just feeling how well it's working out. And if it's not working out, we adjust our grip." Hillis speaks of this rapid-fire feedback mechanism as a kind of "controlled hallucination": we have a hypothesis—a hallucination—about the real world (say, the position of the water glass), sensory feedback from our fingers causes us to adjust the hypothesis, our fingers provide feedback about the validity of the adjusted hypothesis, and so on, until we confidently pick up the glass.

Five years ago, when Hillis was a graduate student, he built a primitive feedback mechanism into a robotic fingertip made up of 256 tiny pressure sensors. The idea was to build a finger that operated by controlled hallucination. The finger could distinguish by touch six different objects—all commonly used fasteners—nuts, bolts, washers, dowel pins, cotter pins, and setscrews. The finger had a "hallucination" about what it was feeling (say, a washer), and then it would test that hallucination (by, say, feeling for the hole in the washer's center). This approach worked well enough in the machine's limited world, but give it anything else, say, a wad of chewing gum, and the finger would confidently identify it as one of the fasteners.

Built with Minsky's guidance, the finger was Hillis's master's thesis and his first foray into parallel processing. Six processors, each as powerful as an IBM personal computer, provided the computational might behind the finger. From this experience, Hillis learned that an awful lot of computational power would be required to bring the repertoire of identifiable objects up to the level of the human forefinger. Hillis is not a man who proceeds cautiously; the next time he hooked together microprocessors, it was a prototype of the Connection Machine.

The decision to connect tens of thousands of processors,

each weaker than a microchip in a video game, came from Hillis's pondering the difference between electronic components in a computer and neurons (nerve cells) in the human brain. The neurons are a million times slower, and yet the brain is far faster than any machine in doing something as simple as distinguishing a man from a woman, reading handwritten characters, or naming a four-letter flower that rhymes with *hose*. Little is known about how the brain does these things, but its blinding speed undoubtedly stems from its having many more basic components than a machine, some hundred billion neurons, give or take a factor of ten. Moreover, says Hillis, "the architecture of the brain, as far as we can see, is completely different from a conventional computer in that it uses a lot of things working in parallel. So that was the intuition behind building a lot of parallel architecture into the Connection Machine."

Hillis is the first to admit that the analogy between the brain and the Connection Machine should not be pushed too far. To begin with, the connections between neurons are on the order of perhaps a hundred trillion, which means a wiring diagram of them is out of the question. Indeed, the connections are so numerous and intertwined that neurobiologists have not yet succeeded in mapping a single neuron, let alone all of them. Therefore, the brain does not provide a model of how to link up the processors in the Connection Machine. Nonetheless, massive parallelism is such a fundamental feature of the brain that it seemed worthwhile trying to make a computer architecture that was similar, even if only remotely and inexactly.

What's more, Hillis recognized how massive parallelism might enable computers to do a host of things, such as image analysis and recognition, that people do effortlessly but that single-processor computers can barely begin to do. No machine, for example, can yet tell a dog from a cat. Conventional computers are handicapped because they must analyze a picture point by point. All the points are stored in the computer's memory and are extracted one at a time along the single, narrow pathway—"the von Neumann bottleneck"—linking the memory and the processor. "It's kind of like scanning a picture by running a peephole over it," says Hillis, rather than processing the entire image at once, as the human vision system does. The Connection Machine has the promise of faring better because each processor is assigned, in effect, to one point in the image and the 65,536 processors, by working together, can analyze the image in its entirety.

The Connection Machine began as Hillis's Ph.D. thesis at MIT. "There was a point when Danny was getting serious about the design," recalls Minsky. "And I said to him, 'I hope you don't make the ILLIAC IV mistake.' And he said, 'Oh, what's the ILLIAC IV mistake?' And I told him."

The ILLIAC IV was a gigantic machine built in the 1970s at the University of Illinois. It had sixty-four processors, each as big as an upright piano because it was built before its time. In fact, they had to use a forklift to plug in the units. It took seven or eight years to build, and the machine was obsolete by the time it was finished. The university gave it to NASA, which promised to use it but had difficulty doing so.

That the project was too ambitious, given the available technology, was not what Minsky meant by the ILLIAC IV mistake. He told Hillis that the concept itself was flawed; it was a mistake, he said, to restrict all the sixty-four processors to doing exactly the same thing at the same time, an expanded electronic version of the Olympic event synchronized swimming. Minsky told Hillis that the processors should be able to operate independently.

"About a month later," Minsky recalls, "Danny came back to me and said, 'Well, I've decided to make the ILLIAC IV mistake.' " Hillis told Minsky that the problem was not in what the processors did but in how they communicated. He said that intraprocessor signals got stuck in traffic jams because, among other things, the connecting wires were restricted to two dimensions. Hillis recognized that a richer connection scheme was required, all the more so since he wanted to hook up not 64 processors but 65,536. The signal traffic would be equivalent to a telephone network serving 65,536 customers who placed a quarter of a billion calls per second. That was the chief technical problem Hillis faced in designing the Connection Machine.

In the end, Hillis and his coworkers at Thinking Machines settled on a three-dimensional architecture in which the processors were connected as if they formed a sixteen-dimensional cube. What this means is that each processor, although directly connected to only 16 others, was never more than sixteen steps away from any of the other 65,536 processors. Moreover, the possibility of a traffic jam was reduced because in sixteen dimensions there are numerous possible routes between any two processors.

The method of transmitting messages was also novel. Hillis describes the communication system as being "halfway between the way the postal system works and the way an old-fashioned telephone system works." In the postal system, there's a lot of flexibility in the route a letter takes. "If a mail plane is full," says Hillis, "they can hold your letter and send it on the next plane." The disadvantage of the postal system is that if you have a lot to say, you must send an awful lot of letters. The advantage of the phone system is that you can stay on the line until you're done communicating. "When you and

I are talking on the phone, at least on a local exchange, there's a wire allocated to us. But when we're not talking, that wire is still there, using up resources." The Connection Machine gets the best of both systems. "It's as if I mailed you a letter with a piece of string attached to it leading to another letter and a string from that letter leading to a third letter and so on. Then we could have constant communication. But if we had to, we could cut the string"—and send the remaining letters by a different route.

When Tom Kraay's superiors at Perkin-Elmer told him to look into the Connection Machine, he was somewhat skeptical. "I couldn't conceive of how you would control 65,000 processors," Kraay recalls. "It sounded blue sky, but it turned out to be easy. No matter what problem I constructed, it ran faster. It's remarkable how much mileage you can get from massive parallelism by splitting up a problem's data so that small chunks are assigned to each processor."

The Connection Machine has not been around long enough for us to see what contributions it will make to artificial intelligence. But it has already proved useful in less exotic fields by making short work of some routine but knotty problems involving document retrieval, circuit design, and air-flow modeling.

Document retrieval is part of the larger problem of searching through vast quantities of information—what computer scientists call a database—for something in particular. This kind of problem is hardly sexy, but it comes up all the time. Moreover, when the database is huge, a conventional computer is intolerably slow. The Connection Machine turns out to be the Evelyn Wood of the computer world when it comes to scanning, say, a year's worth of *New York Times* articles. Instead of looking at the articles one by one, as a conventional machine would, the Connection Machine reads all of them at once because each is effectively assigned to an individual processor. "Imagine 65,000 people in Yankee stadium," says Hillis. "Each person has a different document. You announce a topic over the public address system, and everyone reads his article to see if it matches." That's what the Connection Machine does, but in three hundredths of a second, hundreds of times faster than any other machine.

The design of electronic circuits is a basic industrial task that taxes single-processor computers. In a chip, thousands of electronic components need to be hooked together. Once the connections among components are established in principle, the components must be laid out so that the length of the connecting wires is minimized and overlap largely avoided. A conventional computer does this slowly by changing the circuit design one component at a time. This task is made to order for the Connection Machine because, with each processor representing a different component, various placements of the components can be easily examined. Indeed, the Connection Machine is designing the chips for its own successor, which is rumored to have a million processors.

One of the most exciting potential applications for the Connection Machine is in simulating air flows, perhaps even in modeling the flow of air over a proposed design for an airplane wing. Computer science is not yet at the point where aeronautical engineers can simulate the design of a new aircraft, or even of a wing, on a supercomputer and conclude with confidence that it works. The mathematical equations describing air flow around a plane or a wing are notoriously difficult to solve, for man or machine. Even the construction of an accurate scale model that performs beautifully in a wind tunnel is no guarantee that the real thing will fly. There's no substitute for building and testing a full-scale prototype.

Stephen Wolfram, a chunky, bespectacled physicist in his late twenties with spaghetti hair, long sideburns, and a craving for ice cream at all hours of the day, is the brains behind the Connection Machine's novel approach to air-flow problems. The MacArthur Foundation certified him a "genius," with one of its prestigious awards, and no one who meets him would disagree with this designation.

Wolfram's idea was not to worry about the complex mathematics, which describes the aggregate behavior of the air, but to concentrate on the individual air particles. Each processor is effectively assigned to one particle. In Wolfram's model, the particles all move with the same speed in one of six directions. A simple rule defines how particles scatter if one collides with another. Although the model is quite primitive, with its restrictions on a particle's speed and direction, it seems promising, evidently because molecules don't actually solve mathematical equations but go in, so to speak, with their elbows held high.

If the model works according to plan, the aeronautical engineer will be able to watch the particles bombard a proposed wing design on the screen of his computer terminal. Recently, Wolfram and Hillis determined that with existing technology it would be possible to add hundreds of thousands more processors to the Connection Machine, until it was the size of a small building. Such a machine would cost $100 billion and consume the energy of the largest power station that now exists. "With that machine," says Wolfram, "you could simulate a complete airplane." Others might find that news depressing, but not Wolfram. He says, "It gives me hope. Because if the technology changes, we might ultimately be able to design a plane."

Optimism is not in short supply at Thinking Machines. Future applications are a familiar topic of conversation in the gourmet lunchroom. Hillis and his coworkers speak of the day

when the Connection Machine solves problems that no conventional computer would dare to tackle, not even slowly. One
such problem is a search not of written documents but of
pictures, each picture assigned to its own processor. "You can
imagine asking the machine, 'In what films did the Three
Stooges appear with Zero Mostel?' or 'In which satellite photos
is there corn?' " says Guy Steele, Jr., a senior scientist at Thinking Machines. "I can say with confidence that it will take
massive parallel power to solve this kind of problem."

Wolfram fantasizes about using the Connection Machine to
simulate all sorts of physical systems, so that mathematical
equations can be entirely dispensed with. Hillis looks forward
to the day when a big cousin of the Connection Machine serves
as a computer power plant for an entire city, supplying computing power to every household and commercial establishment.
Minsky plans to use the Connection Machine to understand
how people think, modeling the human brain by having each
processor simulate small groups of nerve cells.

The company as a whole dreams of exploiting the Connection Machine's multiheaded technology to build a true artificial intelligence, a walking, talking robot every bit as versatile
and obedient as C3PO, the cybernetic helpmate in *Star Wars.*
"The long-range goal of this company is to make a real robot,"
says Hillis. "The home robot will eventually be the major
appliance, as important as the automobile is today. It will do
everything you'd want it to do: clean the house, fetch the
paper, clear the dishes, feed the dog."

At IBM, cautious pragmatism, not optimism, is the rule.
The $50 billion giant is convinced that the future of comput-
ing is in parallel processing, but it isn't sure that harnessing the
power of tens of thousands of weak processors is the way to go.
IBM refuses to comment officially on the Connection Ma-

chine, but its researchers say that too much is made of the machine's neuronlike architecture, that Hillis is more of a publicity hound than a scientist, and that, even with his novel communication scheme, he may have underestimated the likelihood of traffic jams.

"I wish we had a better theoretical understanding of the issues involved in parallel processing," says Abe Peled, IBM's vice-president of systems and director of computer science. To get that understanding, IBM, normally a secretive loner, is collaborating with at least eight universities whose researchers are pursuing different strategies. "We can't do it alone," says the IBM computer scientist Greg Pfister, "so our approach is to put out some water and fertilizer, try to get some flowers to grow, and see which one makes it."

At IBM itself, two flowers—RP3 and GF11—are expected to bloom in 1987. RP3 is a forty-person effort, headed by Pfister, to build a machine that contains 512 fairly powerful processors; unlike the more numerous but weaker processors in the Connection Machine, they will not all be restricted to doing the same thing at the same time but can go their own, separate ways. How this greater flexibility will work in practice remains to be seen.

Unlike RP3, which is a jack-of-all-trades computer, the GF11 parallel processor has only one purpose in life: to test physicists' most fundamental theory of the nature of matter. For more than twenty-four hundred years, since the time of the pre-Socratic atomistic philosophers, man has searched for the indivisible constituents of matter. In the past few decades, experimentalists have built powerful machines for smashing bits of matter together with more and more force; in the debris of these collisions, they have identified more than two hundred new species of subatomic particles, exotic relatives of the proton and the neutron, the two familiar constituents of the core of atoms. Many of these particles, in turn, showed signs of an inner structure. In 1964, theorists started to bring order to this untidy zoo of particles by hypothesizing that each structured species was made up of a different combination of a few elementary building blocks called quarks. In the early 1970s, the quark hypothesis became the basis of quantum chromodynamics, or QCD, a comprehensive theory that describes what particles are made of and how they interact.

"Today, QCD looks right in 55 million ways" says Don Weingarten, a particle physicist on the GF11 project, "but all these ways are crude and qualitative. What you'd like is a sharp numerical prediction that you can go out and compare with the result of an experiment." QCD does make sharp numerical predictions about exotic phenomena well beyond the scope of experimental physics, such as a speedy proton as massive as the entire universe. To extract predictions from QCD about down-to-earth things like protons in ordinary matter requires a thorny series of calculations that no man—or existing machine—could ever make. For example, calculating the mass of a stationary proton calls for roughly a hundred quadrillion calculations. That many calculations would tie up a Cray 1 for thirty years and an IBM personal computer for at least two hundred thousand years. The GF11 parallel processor will take only four months, and the result, when compared with the measured mass, may afford the first precise test of quantum chromodynamics.

GF11 certainly doesn't look like a machine at the forefront of theoretical physics. Cooled by two refrigerator-size air conditioners, it takes up a whole room. Four hundred thousand chips forming 576 processors are crammed into twenty huge blue cabinets that resemble oversize gym lockers. The processors are linked by two hundred miles of wire that make up two thousand multicolored snaking cables known around IBM as Monty's Pythons, after Monty Denneau, one of GF11's three designers. "The guy who had to thread all the wires still has a glazed look," says Denneau. "He used to be a musician. Now he can't do anything."

Each of the 576 processors can communicate with any other through an ultrapowerful switching network called the Memphis switch. All messages, even ones between processors that are in the same cabinet, have to be routed through the Memphis switch—just as in the early days of the courier service Federal Express, all packages (even one, say, from Chicago to Detroit) had to be routed through the city of Memphis. The switch can handle the equivalent of two million simultaneous phone conversations.

Like the 65,536 processors in the Connection Machine, the 256 processors in GF11 all execute the same instructions at the same time, but each is much more powerful. Each can multiply a seven-digit number by a seven-digit number in sixty trillionths of a second, making it three thousand times faster than a Connection Machine processor. This kind of brute number crunching is needed to compute the mass of the proton. Although GF11 is designed to do calculations in quantum chromodynamics, IBM hopes that with minor adjustments the machine will be able to tackle other scientific problems that are too computationally intense for conventional machines.

As GF11 and RP3 are being built, the Connection Machine is hard at work—its tiny red lights blinking hypnotically—at Perkin-Elmer's Oakton think tank. The Strategic Defense Initiative requires mirrors to bounce laser beams through space. At Perkin-Elmer, the mirrors are being modeled on the Connection Machine.

Artificial intelligence or not, the Connection Machine looks good for problems in battlefield management. "Today, a battle might involve 100,000 units, each moving at 25,000 miles an hour," says Perkin-Elmer's Tom Kraay. "It will all be over in fifteen minutes, so you'll have to quickly keep track of a lot of things—which is where a Connection Machine would come in."

Perkin-Elmer, Thinking Machine's first commercial customer, could not be more delighted with its purchase. "We love it," says Kraay, "even if others are not yet prepared to discard forty years of wisdom about programming single processors. Parallel processing is like rock and roll. 'What's this yucky stuff? What's this Elvis Presley?' people said at first. But it took over, and so will this."

IV

'ONE MAN, ONE VOTE'

That mathematics is involved in computers is hardly surprising. After all, computers are at bottom just manipulators of o's and 1's. And it was mathematicians like Alan Turing and John von Neumann who designed the first electronic computers.

Long before people ever dreamed of computers, philosophers and political scientists were wrestling with the mechanics of setting up a democratic nation. Here, mathematics rears its ugly head in a surprising and disturbing way. The Nobel Prize–winning work of the American economist Kenneth Arrow shows that achieving the ideals of a perfect democracy is a mathematical impossibility. Indeed, undesirable paradoxes can arise not only in voting but even before voting takes place, in deciding how many representatives are allocated to each district in a system of indirect representation, such as that of the House of Representatives.

12

IS DEMOCRACY MATHEMATICALLY UNSOUND?

Game theory, the mathematical analysis of conflict, be it in politics, business, military affairs, or what have you, was born in 1927, with John von Neumann, the mathematical jack-of-all-trades. Von Neumann recognized that certain decision-making situations in economics and politics are mathematically equivalent to certain games of strategy. Consequently, lessons learned from analyzing these games are directly applicable to decision-making situations in real life. Game theory, also called the science of conflict, was not widely known until 1944, when von Neumann teamed up with Oskar Morgenstern, a Princeton economist, to publish a now-classic book, *Theory of Games and Economic Behavior.*

Part of the intellectual appeal of game theory is that many of its results, like those of quantum mechanics or the theory of relativity, seem counterintuitive, even subversive. Typical is a problem from a 1948 *American Mathematical Monthly* that still comes up in the literature from time to time. Three men, call them Al, Ben, and Charlie, engage in a novel dart game in which balloons are targets. Each contestant has one balloon and remains in the game as long as his balloon is unbroken. The winner is the player who is left with the sole surviving balloon.

Each round, the contestants who remain in the game draw lots to determine the order of play and then take turns throwing one dart apiece. They are all aware of their respective skill: Al can pop a balloon 4 out of 5 times (or 80 percent of the time), Ben can pop one 3 out of 5 times (60 percent of the time), and Charlie can pop one 2 out of 5 times (40 percent of the time). What strategy should each contestant adopt?

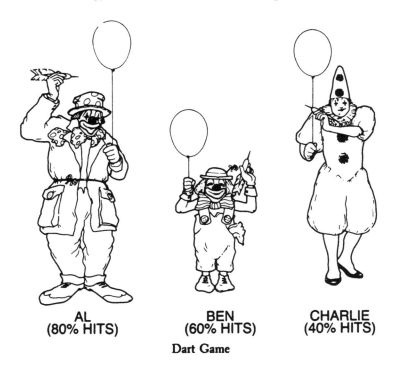

AL
(80% HITS)

BEN
(60% HITS)

CHARLIE
(40% HITS)

Dart Game

The answer seems obvious. Each player should aim at the balloon of the stronger opponent because if he hits it, he'll be left to face only the weaker shot. Nevertheless, if all three contestants follow this sensible-sounding strategy, they finish in reverse order of skill! Probability calculations reveal that Charlie, the worst shot, has the best chance of winning (37 percent) and that Al, the best shot, has the least chance, 30 percent, to Ben's 33 percent.

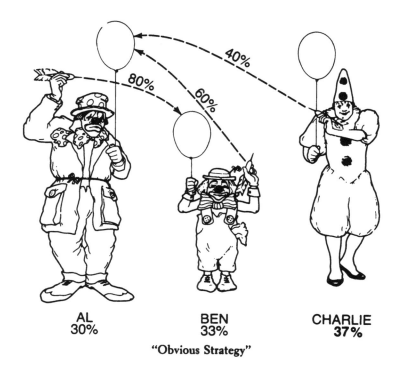

AL
30%

BEN
33%

CHARLIE
37%

"Obvious Strategy"

What has gone wrong? The problem is that while Al and Ben slug it out among themselves, Charlie faces no threat at all. His survivability is enhanced by Al's and Ben's insistence on first doing each other in.

A better strategy for both Al and Ben is not to fire at each other until they have taken out Charlie. Charlie's best counter-strategy is still to throw darts at Al, the stronger opponent. In this case, Al's and Ben's chances of winning improve to 44.4 percent and 46.5 percent, respectively, and Charlie's chances decline dramatically, to 9.1 percent. This scenario may be

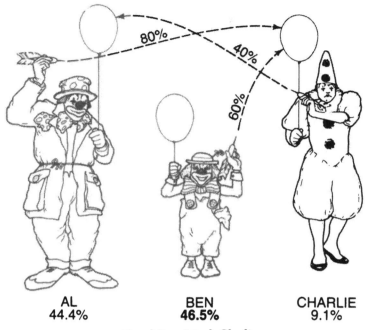

AL
44.4%

BEN
46.5%

CHARLIE
9.1%

Al and Ben Attack Charlie

unstable, however, because it requires Al and Ben to cooperate. Although Al is the best shot, he still doesn't have the best chances of winning, and he might be tempted to double-cross Ben. But if he fails to knock Ben off with the double-cross dart, Ben could fire back—and the calculus of chances of winning would change again.

Instead of cooperating with Ben, whether or not he double-crosses him, Al might try another strategy, discussed in the book *Game Theory in the Social Sciences: Concepts and Solutions,* by Martin Shubik, professor of mathematical institutional economics at Yale University. The idea is that Al, by making verbal threats, tries to set up a situation in which Ben and he are slugging it out but in which Charlie is firing not at him, as in the first scenario, but at Ben. Al announces that he will never throw darts at Charlie's balloon (and always fire at Ben) as long as Charlie never fires at him. Al makes it clear that if Charlie does fire at him, he will fire back. Given the threat of retaliation, probability calculations show that Charlie does best to fire only at Ben's balloon. If Ben attacks Al, the resulting winning chances are 44.4 percent for Al, 20.0 percent for Ben, and 35.6 percent for Charlie. Al has not increased his chances of winning—the percentage hasn't changed—but he's now the front-runner.

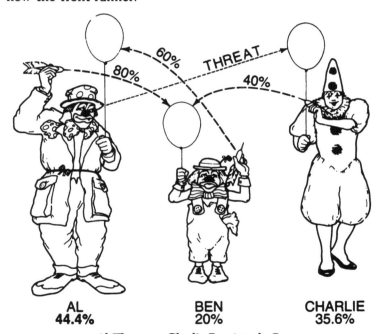

**AL
44.4%**   **BEN
20%**   **CHARLIE
35.6%**

Al Threatens Charlie But Attacks Ben

Naturally, Ben does not want to be the favorite to lose, and so he, like Al, also warns Charlie, "I will not fire at you unless you fire at me, in which case I'll retaliate." Faced with threats from both opponents, Charlie's best strategy is to fire not at either of them but into the air, assuming the rules allow such pacifism!

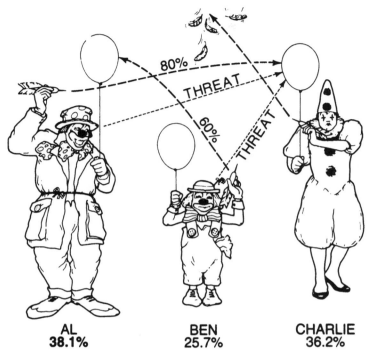

AL
38.1%

BEN
25.7%

CHARLIE
36.2%

Al and Ben Threaten Charlie But Attack Each Other

This curious strategy is best for Charlie, Shubik explains, because his only goal in the first phase of the game, so long as no one is attacking him, is to increase his chances of facing Ben instead of Al in the second phase, the one-on-one encounter. Charlie's cleverness has increased his chances of winning by 0.6 percent, the chances now being 38.1 percent for Al, 25.7 percent for Ben, and 36.2 percent for Charlie. But this is not the final word. Things would get curiouser and curiouser if Al expanded his threat to discourage Charlie from firing into the air.

This problem is typical of many in game theory. The underlying assumption is that each player is rational and that each is out for his own self-interest. One moral of this problem is that the obvious strategy—for each contestant to try to knock off the stronger opponent—is not always good. This is what I mean by a solution's being counterintuitive. Of course, as you plunge further into game theory, your intuitions change and the unexpected is less unexpected, if it's unexpected at all. Another moral of the balloon battle is that the possible solutions cannot be properly evaluated in the absence of information about whether the players can communicate, collude, make threats, and enter into agreements that are binding and enforceable. In game theory, it is often necessary to understand such sociological factors.

Without attempting to be rigorous, we can easily appreciate that the balloon battle might be analogous to a competitive situation in politics or economics. According to Steven Brams, professor of politics at New York University, one of the lessons of the balloon battle might be extended to a multicandidate political race, such as the 1984 Democratic presidential primary in New Hampshire, which had eight candidates. "It might seem," says Brams, "that a candidate's best strategy is to go after the strongest opponent in his part of the political spectrum. If you're a liberal and there are two other liberals, you go after the stronger one. What happens is that the two strongest will shoot each other out, and the weakest liberal will be left." Now, if that happens across the board, the weakest candidate in each segment of the political spectrum will survive—and "there's no way," says Brams, "that a strong candidate will emerge from that kind of field."

In 1951, Kenneth Arrow, an American economist, astounded mathematicians and economists alike with a convincing demonstration that any conceivable democratic voting system can yield undemocratic results. Arrow's unsettling game-theoretic demonstration drew immediate comment in academic circles the world over.

One year later, in 1952, Paul Samuelson, later the winner of the Nobel Memorial Prize in Economic Sciences, put it this way: "The search of the great minds of recorded history for the perfect democracy, it turns out, is the search for a chimera, for a logical self-contradiction. Now scholars all over the world—in mathematics, politics, philosophy, and economics—are trying to salvage what can be salvaged from Arrow's devastating discovery that is to mathematical politics what Kurt Gödel's 1931 impossibility-of-proving-consistency theorem is to mathematical logic."

Arrow's demonstration, called the impossibility theorem (since it showed, in effect, that perfect democracy is impossible), helped earn him the Nobel Prize in economics in 1972. Today, the fallout from Arrow's "devastating discovery," one of the earliest and most astonishing results in game theory, is still being felt.

The undemocratic paradoxes inherent in democratic voting are best explained by an example. Consider three friends, Ronald, Clara, and Herb, who, after a hard day of work, have a craving for fast food. They are determined to dine together at one of three eateries, McDonald's, Burger King, or Wendy's, but they cannot agree on which. Ronald, who longs for a McD.L.T., served in the nifty partitioned container that keeps the greasy hamburger runoff from flooding the crisp, farm-fresh veggies, wants to go to McDonald's; of the other two restaurants, he favors Burger King over Wendy's. Eager to go where the beef is, Clara prefers Wendy's to McDonald's and McDonald's to Burger King. Herb, dreaming of a double Whopper with cheese, likes Burger King best and McDonald's least.

Ronald	Clara	Herb
1. McDonald's	1. Wendy's	1. Burger King
2. Burger King	2. McDonald's	2. Wendy's
3. Wendy's	3. Burger King	3. McDonald's

Fast-Food Preferences

The three friends decide to settle the matter by voting first between McDonald's and Wendy's and then between the winner of that vote and Burger King. If Ronald, Clara, and Herb each vote their real preference, they'll end up at Burger King (with Wendy's the runner-up).

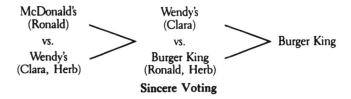

Sincere Voting

Since Burger King is Clara's last choice, she will not be happy. If Clara votes on the first ballot not for her real preference, Wendy's, but for her second choice, McDonald's, she ensures that McDonald's will win the first ballot as well as the second. It is paradoxical that Clara ultimately achieves a preferable result by initially violating her own preference.

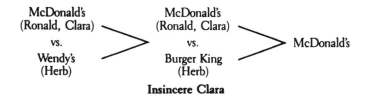

Insincere Clara

Moreover, even if Ronald and Herb get wind of Clara's strategy, they cannot effectively interfere. Herb is outraged because Clara's crafty voting has made his third-choice restaurant the winner, whereas "honest" voting on her part would have made his first choice the winner. Herb tries to persuade Ronald to conspire with him in some insincere voting of their own, but Ronald wants no part of it, because he cannot possibly improve his own position: Clara's voting has made Ronald's first-choice restaurant the winner.

A change in the voting sequence cannot eliminate the possibility of crafty voting. All it would do is give someone other than Clara the opportunity to be insincere. Suppose the three friends vote first between Burger King and Wendy's, with the winner set against McDonald's: if they all vote "honestly," they'll end up at McDonald's, leaving Herb disappointed.

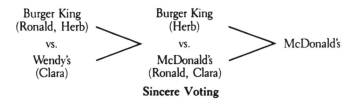

Sincere Voting

If Herb's shrewd enough to foresee this, he'll cast a sly first vote that will steer them ultimately to Wendy's.

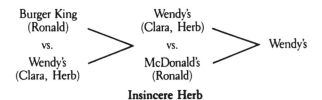

Insincere Herb

The other possible voting sequence—first between McDonald's and Burger King and then between the winner of that vote and Wendy's—is no better.

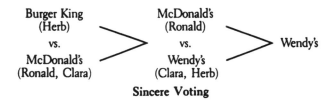

Sincere Voting

It simply gives Ronald the opportunity for shrewd voting.

Insincere Ronald

Although the predicament of the three would-be diners is fictitious, it is not contrived. The possibility of crafty voting may arise in any majority-rule voting in which a series of ballots are cast to select a single winner from three or more alternatives. This happens in the House of Representatives when an amendment to a bill is introduced. First, the House votes on the amendment. If it passes, a second and final vote is taken between the amended bill and the option of no bill at all. If the amendment is defeated, the second vote is between the original bill and no bill.

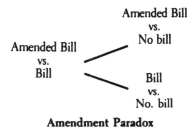

Amendment Paradox

In *Mathematical Applications of Political Science,* William Riker, of the University of Rochester, analyzes a 1956 House vote on a bill calling for federal aid for school construction. An amendment was offered that would provide federal aid only to states whose schools were integrated. The House was essentially divided into three interest groups: Republicans, northern Democrats, and southern Democrats. The Republicans, being against federal aid but in favor of integration, favored no bill at all but preferred the amended bill to the original. The northern Democrats favored the amended bill but preferred the original to no bill. The southern Democrats, being from states with segregated schools, favored the original bill but preferred no bill to the amended bill.

Republicans	Northern Democrats	Southern Democrats
1. No bill	1. Amended bill	1. Bill
2. Amended bill	2. Bill	2. No bill
3. Bill	3. No bill	3. Amended bill

Preferences on School Aid

On the ballot on the amendment, the Republicans provided the winning votes, with the northern Democrats. But on the second vote, between the amended bill and no bill, the Republicans joined forces with the southern Democrats to defeat the amended bill. The paradox here is that in the absence of the amendment, in a straight vote between the original bill and no bill, the original bill would undoubtedly have won!

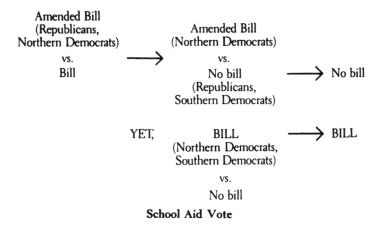

School Aid Vote

"As if it were not enough that the choice may depend on the voting order," Riker concludes, "this fact can be used to twist the outcome of legislative processes. It may be possible to create a voting paradox such that no action is taken by the legislature even though a proposed bill would have passed prior to the creation of the paradox. A legislator could introduce an amendment to create such a paradox, and if the voting order were just right, the amended proposal would then be defeated."

As early as the eighteenth century, the French mathematician Jean-Antoine-Nicolas Caritat, marquis de Condorcet, identified a fundamental voting paradox. He discovered that society often has preferences that, if held by an individual, would be dismissed as irrational. Consider again our three hungry friends. Ronald prefers McDonald's to Burger King and Burger King to Wendy's. Given those preferences, he would be irrational to prefer Wendy's to McDonald's. Yet these are precisely the preferences of our friends as a group! In a vote of all three friends, they prefer McDonald's to Burger King, Burger King to Wendy's, and Wendy's to McDonald's.

Could it be that from a mathematical point of view democracy is inherently irrational?

RONALD'S PREFERENCES:

McDONALD'S ———→ BURGER KING ———→ WENDY'S

THEREFORE McDONALD'S ———→ WENDY'S

SOCIETY'S PREFERENCES:

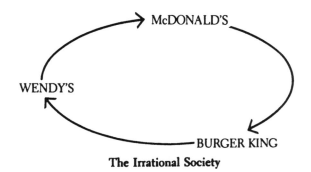

The Irrational Society

The mathematical paradoxes that arise in democratic voting have been extensively studied by Steven Brams, one of the world's most prolific game theorists. He has brought mathematics to bear not only on questions involving voting but also on all sorts of questions that hardly seem to be quantitative. In his book *The Presidential Election Game*, Brams uses game theory to analyze the behavior of Richard Nixon and the Supreme Court in the landmark case that forced the president to release the incriminating "White House tapes." In *Biblical Games*, he applies game theory to conflicts in the Old Testament between God and human beings and concludes that He is a suberb strategist, a touchy, brooding, arbitrary Being who is obsessed with His reputation on Earth. *In Superior Beings: If They Exist, How Would We Know?*, he discusses game-theoretic implications of omniscience, omnipotence, immortality, and incomprehensibility. Brams also applies game theory to down-to-earth subjects, everything from superpower conflicts and professional sports drafts to labor-management negotiations and the scheduling of television shows.

Brams's interest in applied mathematics goes back to his undergraduate days, just after Sputnik, at MIT. He had intended to major in physics but scrapped the idea when he discovered that he was a total klutz in the laboratory. Leaving broken equipment in his wake, he turned to mathematics, at which he had always excelled. He also took courses from big-name professors in the relatively new political-science department. There he found his metier: applying mathematics to political situations. His first efforts were chiefly statistical, involving mathematical modeling of international trade flow. From MIT, he went to Northwestern to do graduate work because it had an avant-garde, very quantitative program in political science.

"Like every self-respecting political scientist," says Brams, "I figured that I should do some work in government, but it wasn't the Peace Corps for me." During the summers of 1963 and 1964, he worked first for the director of the National Institutes of Health and then for the Office of the Secretary of Defense. When he completed graduate school, he went to work full-time at the Institute for Defense Analysis (IDA), a nonprofit research organization that did most of its work for the Joint Chiefs of Staff and the Office of the Secretary of Defense. "I was specifically hired to do a study of how decisions were made in the Department of Defense," recalls Brams. "For six months, I designed and pretested a questionnaire. I was about to go into the field and interview high-level people—assistant secretaries, generals, admirals—when the president of

IDA stopped the study. The Vietnam War was heating up then, and he felt that the study was too controversial, particularly since the Department of Defense was IDA's chief client. I was very chagrined and decided that the only place I'd have the freedom and independence to do what I wanted was at a university."

He started teaching at the University of Rochester, which had the most quantitative political-science department in the country. Riker, who analyzed the 1956 House vote on school construction, was at Rochester, and Brams picked up from him a fascination with game theory. "And," says Brams, "I've never been off the subject since."

Critics of game theory sometimes charge that it is an insidious discipline, a mathematical stamp of approval for the crafty tactics of political operators. Game theory, however, does not create voting paradoxes; it merely identifies them in a formal way. The paradoxical vote in the House on the 1956 school construction bill came about naturally, not because congressmen took their cue from some Machiavellian game-theoretic journal.

Once a paradox is formally identified, game theory can help evaluate how often it arises. Consider, for example, Condorcet's observation that society's preferences, as determined by a vote of individuals, may be paradoxically "intransitive": that society could prefer McDonald's to Burger King and Burger King to Wendy's but favor Wendy's over McDonald's. If society consists of three individuals (Ronald, Clara, and Herb), the intransitivity will come about only when each of the restaurants is ranked first by one individual, second by another, and third by the other. Given that all possible individual preferences are equally likely, the chance of a societal intransitivity is 5.6 percent. That may not seem like much, but keep in mind that this percentage is only for the simplest case of three people and three alternatives.

In the book *Paradoxes in Politics,* Brams summarizes recent studies of the probability of societal intransitivity in more complex cases. It turns out that the probability increases both as the number of alternatives increases and as the number of voters increases but that it is more sensitive to the number of alternatives. If the number of alternatives is held fixed at three, the possibility of the paradox increases slightly, from 5.6 percent (for three voters) to 8.8 percent (as the number of voters approaches infinity). If the number of voters is held fixed at three, the possibility of the paradox skyrockets, from 5.6 percent (for three alternatives) to 100 percent (as the number of alternatives approaches infinity). Indeed, notes Brams, for any fixed number of voters, the probability of the paradox climbs to certainty as the number of alternatives increases indefinitely.

	NUMBER OF VOTERS					
	3	5	7	9	11	∞
3	5.6%	6.9%	7.5%	7.8%	8.0%	8.0%
4	11.1%	13.9%	15.0%	15.6%	16.0%	17.6%
5	16.0%	20.0%	21.5%	23.0%	25.1%	25.1%
6	20.2%	25.5%	25.8%	28.4%	29.4%	31.5%
7	23.9%	29.9%	30.5%	34.2%	34.3%	36.9%
∞	100.0%	100.0%	100.0%	100.0%	100.0%	100.0%

NUMBER OF ALTERNATIVES (row labels at left)

Chance of Societal Intransitivity

From Steven Brams, *Paradoxes in Politics* (New York: Free Press, 1976), 42.

The mathematics in game theory may be simple compared with what is involved in many other abstract mathematical disciplines, but it is certainly not trivial. Indeed, the mathematics often leads to results that counter intuition or defy expectation. The simplicity of the mathematics does not make game theory any less rigorous than, say, higher-dimensional topology, in which the journals can be read only by a handful of Ph.D.'s. The simplicity may even be a virtue: the mathematics in game theory is so accessible that there is little likelihood that mathematically dubious statements will go unnoticed in the literature.

One such statement, which Brams exposed as being incorrect, was made by—of all people—the officers of the American Mathematical Society. That such a distinguished group of mathematicians could slip up shows how surprising the results of game theory can be. The erroneous statement appeared on the instructions for a ballot that AMS members were using to elect representatives to a special committee. For the balloting, the AMS resurrected a voting procedure, the system of single transferable vote (also called preferential voting), that was advocated in the late 1850s by Thomas Hare, an obscure English barrister who wrote two books critical of traditional voting systems.

Hare was particularly distressed by the fact that in traditional systems of proportional representation in which each voting district elects more than one candidate, sizable minorities may effectively be disenfranchised, even though their raw numbers suggest that they are entitled to representation. Consider an imaginary district in which two of four candidates are to be elected. Two of the candidates, call them Attila the Hun and GI Joe, are card-carrying conservatives; of the two, Attila is more right wing. The other two candidates, Hal Handout and Freeda Freelove, are liberals; of the two, Freelove's heart bleeds more profusely. There are twenty-three voters in the district, of which thirteen are conservative and ten are liberal. The preferences of the twenty-three, in ranking the candidates from first choice to last choice, are as follows:

Number of Voters	First Choice	Second Choice	Third Choice	Fourth Choice
7	Attila	GI Joe	Handout	Freelove
6	GI Joe	Attila	Handout	Freelove
6	Handout	Freelove	GI Joe	Attila
4	Freelove	Handout	GI Joe	Attila

In an election in which each voter is allowed to vote for two candidates, Attila and GI Joe will win, with thirteen votes apiece. As a result, the ten liberal voters will not be represented, even though they constitute 43 percent of the electorate. The thirteen conservative voters, who make up 57 percent of the electorate, will have 100 percent of the representation.

Believing that the chosen representatives should mirror more closely the makeup of the electorate, Hare devised an ingeniously complex system that requires each voter to list the candidates in order of preference, so far as he can distinguish among them. The first-choice votes are then tabulated, and the candidates who achieve a certain quota of votes are the winners.

The quota is computed to be the minimum number of first-place votes such that the maximum number of candidates who could meet the quota corresponds to the number of open seats. In the example above, with twenty-three voters and two open seats, the quota is eight; two candidates (but not three) could get eight first-place votes. A quota of seven would be too low because three candidates could conceivably meet the quota, one candidate too many, since only two seats are open. (In general, the quota can be found by dividing the number of voters by one more than the number of open seats, adding one to this quantity, and discarding any resulting fraction.)

Assuming that at least one candidate meets the quota and at least one seat remains open, the winning votes in excess of the quota are proportionally transferred to the next-highest choice of those voters. If this transfer causes another candidate to meet the quota, he is elected; and if seats remain unfilled, surplus votes are again proportionally transferred. This process continues until all the seats are filled. If at any point there is an open seat but no surplus votes to transfer, the candidate with the lowest number of votes is eliminated and his supporters simply transfer their votes to their next-highest choice who's still in the race. The idea is that no vote should be wasted: if a vote is more than a candidate needs to be elected, it should count elsewhere; if it's squandered on the least-popular candidate, it should also count elsewhere.

The best way to understand these rules is to apply them to a specific example. Let's try them in our imaginary district. Since the quota is eight, each of the four candidates falls short. Consequently, the least-popular candidate, Freeda Freelove, is eliminated, and her four supporters transfer their votes to Hal Handout, their second choice. If Freelove is scratched from the preference lists, the tally is as follows:

Number of Voters	Preferences (Best to Worst)		
7	Attila	GI Joe	Handout
6	GI Joe	Attila	Handout
10	Handout	GI Joe	Attila

Now, Hal Handout has exceeded the quota by two votes, so he is elected. His two surplus votes are passed to GI Joe:

Number of Voters	Preferences (Best to Worst)	
7	Attila	GI Joe
8	GI Joe	Attila

This time around GI Joe has reached the quota, so he wins the other seat.

The election of Handout and Joe would please Hare: both the conservatives and liberals are represented, and the more extreme candidates in each camp failed to win. Impressed by results like this one, John Stuart Mill praised Hare's system "as among the greatest improvements yet made in the theory and practice of government." Today, Hare's system is used in legislative elections in Australia, Malta, the Republic of Ireland, and Northern Ireland, school board elections in New York City, and city council elections in Cambridge, Massachusetts, not to mention in balloting in many professional organizations like the American Mathematical Society.

The AMS ballot included two strong statements: "There is no tactical advantage to be gained by marking fewer candidates" and "It is advisable to mark candidates in the order of your preference until you are ignorant or indifferent concerning candidates whom you have not ranked." Brams constructed an example to show that this is not true, that it may be advantageous to mark fewer candidates. Suppose there are seventeen voters, two open seats, and four candidates, whom I've named Dr. Graph, Dr. Digit, Dr. Point, and Dr. Manifold. The preferences of the voters are as follows:

	Number				
Class	*of Voters*		*Preference Order (Best to Worst)*		
A	6	Dr. Graph	Dr. Digit	Dr. Point	Dr. Manifold
B	6	Dr. Graph	Dr. Point	Dr. Manifold	Dr. Digit
C	5	Dr. Graph	Dr. Manifold	Dr. Digit	Dr. Point

Dr. Graph wins with seventeen votes, a surplus of eleven votes above the quota of six. Therefore, eleven votes need to be transferred. In a situation like this, where the voters who support the winner do not agree on their other choices, Hare's rules (which the AMS follows) require that the eleven surplus votes be proportionally transferred: $6/17$ of eleven votes going to class A, $6/17$ of eleven votes going to class B, and $5/17$ of eleven votes going to class C. The result is the following:

	Number			
Class	*of Voters*		*Preference Order (Best to Worst)*	
A	3.9	Dr. Digit	Dr. Point	Dr. Manifold
B	3.9	Dr. Point	Dr. Manifold	Dr. Digit
C	3.2	Dr. Manifold	Dr. Digit	Dr. Point

Since no one meets the quota, the least-popular candidate, Dr. Manifold, is eliminated, and the 3.2 votes of his supporters are transferred to their next-highest choice, Dr. Digit:

	Number		
Class	*of Voters*	*Preference Order (Best to Worst)*	
A	7.1	Dr. Digit	Dr. Point
B	3.9	Dr. Point	Dr. Digit

Now, Dr. Digit has surpassed the quota of six, so he joins Dr. Graph as an elected candidate.

The six voters in Class B (with the preference order Dr. Graph, Dr. Point, Dr. Manifold, and Dr. Digit) are pleased that their first choice was elected but disturbed that their last choice was also chosen. Suppose the election is repeated and everything is the same except that two of these six voters decide to ignore the AMS's statement ("There is no tactical advantage to be gained by marking few candidates") and mark their ballots only for Dr. Graph. The preferences now break down into four categories:

Class	Number of Voters	Preference Order (Best to Worst)			
A	6	Dr. Graph	Dr. Digit	Dr. Point	Dr. Manifold
B'	4	Dr. Graph	Dr. Point	Dr. Manifold	Dr. Digit
B"	2	Dr. Graph			
C	5	Dr. Graph	Dr. Manifold	Dr. Digit	Dr. Point

Again, Dr. Graph is the unanimous choice on the first tally. The eleven surplus votes of his supporters are distributed 6/17 of eleven to class A, 4/17 of eleven to class B', 2/17 of eleven to class B", and 5/17 of eleven to class C. But class B" drops out because its members did not mark preferences beyond their first choice. Thus the situation is this:

Class	Number of Voters	Preference Order (Best to Worst)		
A	3.9	Dr. Digit	Dr. Point	Dr. Manifold
B'	2.6	Dr. Point	Dr. Manifold	Dr. Digit
C	3.2	Dr. Manifold	Dr. Digit	Dr. Point

As in the first election, a second candidate did not reach the quota, so the lowest vote getter, Dr. Point, is scratched and the 2.6 votes of his supporters are incorporated into class C:

Class	*Number* *of Voters*	*Preference Order (Best to Worst)*	
A	3.9	Dr. Point	Dr. Manifold
C, B′	5.8	Dr. Manifold	Dr. Point

The two remaining candidates are both short of the quota of six, but Dr. Point is eliminated since he has fewer votes, and Dr. Manifold is declared the winner. The two clever voters in class B who marked short ballots ended up with a preferable result: their third choice won a seat instead of their fourth choice.

In an actual election, such a result might be difficult to achieve. "I wish to make clear," Brams writes, "that I am not suggesting that voters would routinely make the strategic calculations implicit [in this counterexample to the AMS's ballot instruction]. These calculations are not only rather complex but also could, on occasion, be neutralized by counterstrategic calculations of other voters in gamelike maneuvers. Rather, I am suggesting that the advice to rank *all* candidates for whom one has preferences is not always rational under the Hare system."

Moreover, the result in Brams's counterexample would be derailed if too many of the voters in class B tried to be clever and cast short ballots. Suppose five of the six voters list only Dr. Graph on their ballots. Then, after the first tally, the situation is as follows:

Class	*Number* *of Voters*	*Preference Order (Best to Worst)*		
A	3.9	Dr. Digit	Dr. Point	Dr. Manifold
B′	0.6	Dr. Point	Dr. Manifold	Dr. Digit
C	3.2	Dr. Manifold	Dr. Digit	Dr. Point

Since no one meets the quota, Dr. Point is forced to drop out, and his supporters join class C:

	Number	
Class	of Voters	Preference Order (Best to Worst)
A	3.9	Dr. Digit Dr. Manifold
C, B'	3.8	Dr. Manifold Dr. Digit

This time Dr. Manifold must bow out, leaving Dr. Digit the victor, as he originally was when the six voters in class B ranked all four candidates on their ballots.

Lest you think that Brams's counterexample turns on the fractions that result when votes are proportionally transferred, he constructs another counterexample in which only whole votes are transferred, as candidates are eliminated. The example involves twenty-one voters who are electing a single candidate from a field of four. Since only one candidate is being chosen, the balloting system is an elimination contest that ends as soon as one candidate has at least the bare majority of eleven votes. I leave it to you to play the role of the game theorist and construct a counterexample. The goal, of course, is to assign preferences in such a way that some voters might benefit by ignoring the AMS's advice. (At the end of this chapter, you'll find the counterexample that Brams came up with.)

The problems with the Hare system run much deeper than these admittedly contrived counterexamples, which can come about only given certain preferences and then only if certain voters have precise information about all the preferences of their fellow voters, if "enemy" voters don't adopt effective counterstrategies, and if too many like-minded voters don't try to be crafty. Gideon Doron and Richard Kronick of the University of Rochester call attention to a perverse feature of the Hare system that can arise even if all the voters cast sincere ballots reflecting their complete preferences.* Under the Hare

*See Doron and Kronick, "Single Transferrable Vote: An Example of a Perverse Social Choice Function," American Journal of Political Science 21 (May 1977): 303–11.

system, Dorin and Kronick note, a candidate may be hurt if he receives additional votes. Indeed, more votes can make a winner a loser!

To understand this perverse possibility, consider Doron and Kronick's example, embellished by our old friends Attila the Hun, GI Joe, Hal Handout, and Freeda Freelove. This time the district has twenty-six voters. Two candidates are to be elected, so the vote quota is nine. The preferences of the twenty-six voters are diverse, not split along liberal-conservative lines:

Class	Number of Votes	Preferences (Best to Worst)			
A	9	Attila	GI Joe	Handout	Freelove
B	6	Handout	Freelove	GI Joe	Attila
C	2	Freelove	Handout	GI Joe	Attila
D	4	Freelove	GI Joe	Handout	Attila
E	5	GI Joe	Handout	Freelove	Attila

Since Attila has reached the quota, he is elected. Attila has no surplus votes, so the lowest vote getter, GI Joe, is eliminated, and his five votes are transferred to class B:

Class	Number of Votes	Preferences (Best to Worst)	
B, E	11	Handout	Freelove
C	2	Freelove	Handout
D	4	Freelove	Handout

Handout, with eleven votes, is therefore elected.

Consider, now, a second set of preferences that is identical to the previous set, with the exception that the two voters who preferred Freelove to Handout (class C) now prefer Handout to Freelove (class C'). In other words, the preferences of class C' are identical to those of class B, so that Handout starts with eight first-place votes, two more than before:

Class	Number of Votes	Preferences (Best to Worst)			
A	9	Attila	GI Joe	Handout	Freelove
B	6	Handout	Freelove	GI Joe	Attila
C'	2	Handout	Freelove	GI Joe	Attila
D	4	Freelove	GI Joe	Handout	Attila
E	5	GI Joe	Handout	Freelove	Attila

Again, Attila is immediately elected and has no surplus votes to transfer. This time, however, the lowest vote getter is Freelove, not GI Joe, and Freelove's four votes are combined with the five votes in class E, pushing GI Joe over the quota:

Class	Number of Votes	Preferences (Best to Worst)	
B	6	Handout	GI Joe
C'	2	Handout	GI Joe
E, D	9	GI Joe	Handout

The outcome could not be more perverse. Recall that all the preference orders are the same, except that two voters elevated Handout from second choice to first. This had the effect of denying him election. "It is simply not fair," Doron and Kronick conclude, "that a candidate could lose an election because he or she received *too* many votes. Most voters would probably be alienated and outraged upon hearing the hypothetical (but theoretically possible) election night report 'Mr. O'Grady did not obtain a seat in today's election, but if 5,000 of the supporters had voted for him in second place instead of first place, he would have won!' "

The perverse possibility that more votes can turn a winner into a loser is not just an artifact of the Hare system. In their book *Approval Voting*, Brams and Peter Fishburn, a mathematician at AT & T Bell Labs, show that it can also plague such familiar voting systems as a plurality election followed by a runoff between the top-two vote getters. Consider three candidates, Marco Denunzio, Patrick O'Rourke, and Basil Jefferson, and seventeen voters who have the following preferences:

Class	Number of Votes	Preferences (Best to Worst)		
A	6	Denunzio	O'Rourke	Jefferson
B	5	Jefferson	Denunzio	O'Rourke
C	4	O'Rourke	Jefferson	Denunzio
D	2	O'Rourke	Denunzio	Jefferson

If all the voters vote sincerely, Denunzio (with six votes) and O'Rourke (also with six) will end up in the runoff, which Denunzio will win, eleven votes to six.

Now imagine that the preferences are the same, except that the last class of voters elevates Denunzio from second choice to first choice:

Class	Number of Votes	Preferences (Best to Worst)		
A	6	Denunzio	O'Rourke	Jefferson
B	5	Jefferson	Denunzio	O'Rourke
C	4	O'Rourke	Jefferson	Denunzio
D'	2	Denunzio	O'Rourke	Jefferson

On the first ballot, Denunzio (eight votes) and Jefferson (five) make the runoff, which Denunzio then loses, eight votes to nine, because O'Rourke's four supporters join Jefferson's. Denunzio's increased support has perversely torpedoed his victory.

Brams also suggests that the public announcement of how candidates fared in a preelection poll might have the same perverse effect in a straightforward plurality election without a runoff. Given the first set of preferences above, in which the two class D voters prefer O'Rourke to Denunzio, the poll results would inform Jefferson's supporters that their candidate was in last place. Jefferson's supporters would have the information they need to abandon their candidate and vote strategically for their second choice, Denunzio, who would thereby win. Given the second set of preferences above, in which Denunzio has picked up the support of the class D voters, the poll results would inform O'Rourke's supporters that their candidate was in last place. Consequently, they would throw their support to Jefferson, who would then beat out Denunzio, in spite of Denunzio's picking up the support of two more voters. In effect, the poll takes the place of a first ballot, making the actual election equivalent to a runoff.

In another paper,* Doron points out another troubling feature of the Hare system: a candidate who wins in two separate districts can lose in a combined tally of the two districts. In Doron's example, a single candidate is to be elected from a group of four. Each district has twenty-one voters, so the quota in each is eleven.

DISTRICT 1

Class	Number of Votes	Preferences (Best to Worst)			
A	8	Attila	GI Joe	Handout	Freelove
B	4	GI Joe	Handout	Freelove	Attila
C	3	Handout	Attila	Freelove	GI Joe
D	6	Freelove	Handout	GI Joe	Attila

*"The Hare Voting System Is Inconsistent," *Political Studies* 27 (June 1979): 283–86.

DISTRICT 2

Class	Number of Votes	Preferences (Best to Worst)			
A	8	Attila	GI Joe	Handout	Freelove
B	4	GI Joe	Handout	Freelove	Attila
C	6	Handout	Attila	Freelove	GI Joe
D'	3	Freelove	Attila	GI Joe	Handout

In both of the districts, no one initially has the quota of eleven. In district 1, Handout is eliminated since he received the fewest first-place votes; the votes of his supporters are transferred to Attila, who wins with eleven votes. In district 2, Attila is also the winner, as he picks up the three votes from Freelove, the lowest vote getter.

Consider what happens when the two districts are merged into a single district, where the preferences of each of the forty-two voters remain the same:

SINGLE DISTRICT

Class	Number of Votes	Preferences (Best to Worst)			
A	16	Attila	GI Joe	Handout	Freelove
B	8	GI Joe	Handout	Freelove	Attila
C	9	Handout	Attila	Freelove	GI Joe
D	6	Freelove	Handout	GI Joe	Attila
D'	3	Freelove	Attila	GI Joe	Handout

The quota is now twenty-two votes. Since voter preferences are exactly the same, it would be perversely inconsistent if Attila were not the victor. But perversity carries the day. Since no one has the quota, GI Joe is eliminated, and the eight votes of his supporters are transferred to their second choice, Handout:

SINGLE DISTRICT

Class	Number of Votes	Preferences (Best to Worst)		
A	16	Attila	Handout	Freelove
B	8	Handout	Freelove	Attila
C	9	Handout	Attila	Freelove
D	6	Freelove	Handout	Attila
D'	3	Freelove	Attila	Handout

Again, all the candidates fall short of the quota, so Freelove, who has the fewest votes, is eliminated. Freelove's three supporters in class D' transfer their votes to their third choice, Attila, whereas Freelove's six supporters in class D transfer theirs to Handout:

SINGLE DISTRICT

Class	Number of Votes	Preferences (Best to Worst)	
A, D'	19	Attila	Handout
B, C, D	23	Handout	Attila

Handout has emerged the winner, with twenty-three votes.

This perverse result can also arise in the reverse direction, when a large district is split into two smaller ones. Forward or reverse, this possibility "makes gerrymandering a very attractive option to affect election results," Doron concludes.

And this is by no means the end of the paradoxes! In an entertaining article,* Brams and Fishburn call attention to two other disturbing features of the Hare system: the no-show

---

*See Fishburn and Brams, "Paradoxes of Preferential Voting," *Mathematics Magazine* 56 (September 1983): 207–14.

paradox and the thwarted-majorities paradox. In the no-show paradox, the addition of ballots on which a certain candidate is ranked last may make that candidate a winner instead of a loser. In other words, voters who rank that candidate last may be better off staying home than filling out a ballot on which they rank him last! In the thwarted-majorities paradox, a certain candidate does not win even though he could beat each of the other candidates in a face-to-face race. (I urge those of you who aspire to be game theorists to construct numerical examples that demonstrate each of these paradoxes; if you do not succeed, you can always consult Brams and Fishburn's very readable paper.)

The thwarted-majorities paradox afflicts not just the bizarre Hare system but also many common voting systems, such as a simple-plurality election. Imagine a three-way race between Mr. Liberal (preferred by 49 percent of the electorate), Mr. Moderate (preferred by 10 percent), and Mr. Conservative (preferred by 41 percent). Now, consider the second choice of each of the three constituencies. The liberal voters naturally prefer Mr. Moderate to Mr. Conservative, so in a two-way contest between these candidates, Mr. Moderate will win, with 59 percent of the vote (to Mr. Conservative's 41 percent). The conservative voters naturally prefer Mr. Moderate to Mr. Liberal, so in a two-way contest between these candidates, Mr. Moderate will win, with 51 percent of the vote (to Mr. Liberal's 49 percent). In a three-way race, however, Mr. Moderate will come in last. In certain primaries, a runoff election is held between the top-two vote getters if no candidate wins a majority. Mr. Moderate would be excluded from such a runoff even though he could beat either opponent in a two-way race.

The paradox may run even deeper. Suppose that on the political spectrum Mr. Liberal is way left of center and Mr.

Conservative is only slightly right of center. In that case, in the runoff between Mr. Liberal and Mr. Conservative, all the moderate votes might go to Mr. Conservative, making him the winner, with 51 percent of the vote. Now we have a curious alignment of preferences in which Mr. Conservative wins on two ballots, Mr. Liberal wins on one ballot, and Mr. Moderate has enough strength to defeat either opponent in a head-to-head contest. You choose your voting system, and you've chosen your winner.

Brams advocates a voting system called approval voting. It either altogether eliminates the paradoxes discussed here, reduces their likelihood, or diminishes their impact. Approval voting replaces the time-honored principle "One man, one vote" with the principle "One man, many votes." In other words, each voter can approve of (that is, vote for) as many candidates as he likes, although he can cast only one vote per candidate. The idea is that a voter need never fear that he is wasting his vote on an unpopular candidate (say, John Anderson in the 1980 presidential election), because he can also vote for whomever else he approves of.

Under approval voting, the winner will not be a candidate who, in a simple-plurality election, ekes out a victory because his opponents split the vote. Approval voting is less likely to thwart the wishes of the majority. And when the majority has no clear-cut preference (in other words, when there is a societal intransitivity, when society prefers McDonald's to Burger King, Burger King to Wendy's, but Wendy's to McDonald's), approval voting will select the choice that meets with the most approval. We saw how, when Ronald, Clara, and Herb voted on restaurants in two ballots, it could be advantageous to vote insincerely, for your second choice instead of your first choice. When there are three candidates, approval voting is immune to this kind of insincere voting: it is never to your advantage to vote for a second choice without also voting for a first choice. Moreover, under approval voting you never benefit from staying home and not voting, as you do in the Hare system, and no funny business happens when districts are combined or split.

Despite these manifest advantages, approval voting is apparently used nowhere in the world in a public forum (other than a few professional societies) except in the United Nations Security Council, where member nations can vote for more than one candidate for the post of secretary general. New York and Vermont thought about using approval voting, but bills to enact it died in the state legislature. The game theorist's role in influencing public policy is a minor one, even when he's come up with a proposal whose benefits to society seem to be mathematically unassailable.

ANSWER TO PROBLEM POSED

Here's Brams's example of a situation in which it could be advantageous to cut short your ballot in the Hare system of voting. There are eleven voters and four candidates for one office.

Class	Number of Votes	Preference Order (Best to Worst)			
A	7	Dr. Graph	Dr. Manifold	Dr. Digit	Dr. Point
B	6	Dr. Manifold	Dr. Graph	Dr. Digit	Dr. Point
C	5	Dr. Digit	Dr. Manifold	Dr. Graph	Dr. Point
D	3	Dr. Point	Dr. Digit	Dr. Manifold	Dr. Graph

Since no candidate has eleven votes, the lowest vote getter, Dr. Point, is scratched, and the three votes of his supporters are transferred to class C:

Class	Number of Votes	Preference Order (Best to Worst)		
A	7	Dr. Graph	Dr. Manifold	Dr. Digit
B	6	Dr. Manifold	Dr. Graph	Dr. Digit
C, D	8	Dr. Digit	Dr. Manifold	Dr. Graph

Still, no one has a simple majority of the votes, so again the least-popular candidate, Dr. Manifold, is eliminated. When the six votes of his supporters are combined with the seven votes in class A, Dr. Graph is elected, with a total of thirteen votes.

The three voters in class D are unhappy because their last choice was the victor. Suppose they had marked only their first choice on the ballot:

Class	Number of Votes	Preference Order (Best to Worst)			
A	7	Dr. Graph	Dr. Manifold	Dr. Digit	Dr. Point
B	6	Dr. Manifold	Dr. Graph	Dr. Digit	Dr. Point
C	5	Dr. Digit	Dr. Manifold	Dr. Graph	Dr. Point
D	3	Dr. Point			

As before, no one initially has eleven votes, and Dr. Point is eliminated. This time, however, his three votes are not transferred, since his supporters did not indicate any other preferences. Of the three remaining candidates, Dr. Digit is now the least popular. When his five votes are incorporated into class B, Dr. Manifold emerges the winner—a result more to the liking of the voters in class D.

13

THE QUANTUM CONGRESS

In 1882, Roger Q. Mills, a Texas congressman, uttered one of the most heartfelt denunciations of mathematics ever to pass through human lips: "I thought that mathematics was a divine science. I thought that mathematics was the only science that spoke to inspiration and was infallible in its utterances. I have been taught always that it demonstrated the truth. I have been told that while in astronomy and philosophy and geometry and all other sciences, there was something left for speculation, that mathematics, like the voice of Revelation, said when it spoke, 'Thus saith the Lord.' But here is a new system of mathematics that demonstrates the truth to be false."

Mills was speaking of a problem that has faced the House of Representatives since the beginning of the Republic: How many representatives should each state be allotted? The mathematics of congressional apportionment may sound like a simple application of the cherished idea of one man, one vote. But, like direct voting schemes, systems of indirect representation are plagued by mathematical paradoxes so devilishly striking that they provoked Congressman Mills to new rhetorical heights. The paradoxes in direct voting schemes are game-theoretic in nature; they involve voters conniving to elect their own candidates. The issue in congressional apportionment is the number of representatives each state is allowed, not how the representatives are elected. Apportionment belongs to an area of applied mathematics called social choice theory.

Why is apportionment such a problem? Article I, Section II, of the Constitution of the United States seems to provide a straightforward solution: The number of representatives each state sends to the House of Representatives shall be proportional to the state's population. The problem is that although the loyalties of a congressman can be divided, his body cannot be; human beings, like pennies or electric charges or subatomic spin states, are quantized.

Suppose you want to set up a House of Representatives in a country that comprises only two states: state X, with a population of 11, and state Y, with a population of 23. What is the smallest house in which each state could be represented according to its population? The smallest house would have 34 members; with fewer members, one of the states (or both of them) would have a fractional number of representatives. In other words; when H (the size of the house) is less than 34, there are no integers X and Y (the numbers of representatives from states X and Y, respectively) that satisfy the equations $X + Y = H$ and $X/Y = 11/23$. And of course a house of 34 for a population of 34 is not exactly indirect representation.

The problem is obviously compounded for a country the size of ours, with fifty states whose populations are not integral multiples of one another. For a house of a given size, the ideal number of representatives for each state is found by multiplying the ratio of the state's population to the total population by the total number of house members. (So, for a house of 235 seats, a state whose population is 2,559,253 in a nation of 231,575,493 would be ideally entitled to 2.597099 representatives: 2,559,253/231,575,493 $\times$ 235.) Since this ideal number will probably be fractional, and a quarter of a representative is not permitted, a better method is needed for allotting the number of representatives.

Many of the Founding Fathers, including Alexander Hamilton, Thomas Jefferson, and Daniel Webster, came up with their own solutions. Treasury Secretary Hamilton's method, which is the easiest to understand, was approved by Congress in 1792 but the vetoed by George Washington—the first presidential veto and one of only two vetoes that Washington exercised in his eight years in office. According to Hamilton's method, each state is initially entitled a number of representatives equal to the integral part of its ideal representation, the fractional part being discarded. In other words, if Vermont is ideally entitled to 3.62 representatives, it gets to have 3 representatives. The number of representatives allotted on this basis is then totaled, and if the total falls short of the designated house size, the house is filled by allocating additional representatives to the states with the largest discarded fractions.

Hamilton's method of apportionment is easy to illustrate. The table below shows the populations of five states and the number of representatives that each would receive in a house of 26 seats.

State	Population	Ideal Number of Reps. in a 26-Seat House	First Round of Hamilton Allotment	Second Round of Hamilton Allotment
A	9,061	9.061	9	9
B	7,179	7.179	7	7
C	5,259	5.259	5	5
D	3,319	3.319	3	4
E	1,182	1.182	1	1
Total	26,000	26	25	26

By Hamilton's method, in a 26-seat house, states A, B, C, D, and E initially receive the following numbers of representatives, respectively: 9, 7, 5, 3, and 1. But that accounts for only 25 of the 26 seats, and state D, having the highest fraction (.319), thus receives an additional representative, for a total of 4.

Hamilton's method always satisfies at least one criterion of equity: it gives each state its ideal number of representatives rounded down or rounded up. In other words, if state D is ideally entitled to 3.319 representatives, Hamilton's method always provides state D with either 3 or 4 representatives, never 2 or 5. A method that adheres to this natural criterion is said to satisfy *quota*. Many methods do not satisfy quota, which seems to be the minimum that you'd expect of an apportionment method that claims to be fair.

Hamilton's method, however, violates another, subtler criterion of fairness. Imagine that the size of the house in our five-state example is increased from 26 to 27:

State	Population	26-SEAT HOUSE		27-SEAT HOUSE	
		Ideal Number	*Hamilton Allotment*	*Ideal Number*	*Hamilton Allotment*
A	9,061	9.061	9	9.410	9
B	7,179	7.179	7	7.455	8
C	5,259	5.259	5	5.461	6
D	3,319	3.319	4	3.447	3
E	1,182	1.182	1	1.227	1
Total	26,000	26	26	27	27

In the 27-seat house, states A, B, C, D, and E receive the following numbers of representatives, respectively: 9, 8, 6, 3, and 1. Amazingly, state D has lost a representative even though the size of the house has increased. This is a serious defect of Hamilton's method. Think of it this way: although neither the total population nor the population of state D changed one iota, state D now has fewer representatives in a larger house. It has been doubly penalized by a cruel quirk of mathematics, called the Alabama paradox (because it was first detected in some calculations involving that state). The five-state example above was concocted by Michael Balinski and H. Peyton Young in a review article on apportionment.* Balinski and Young spent nine years probing the mathematical paradoxes of apportionment and researching the history of political debates on proposed apportionment schemes. Much of my account is based on their work.

This Alabama paradox—that a state can lose representatives in a larger house—was not a factor in Washington's veto of Hamilton's proposal. Indeed, there is no evidence that the Founding Fathers even knew about this mathematical peculiarity. In vetoing Hamilton's proposal, Washington was swayed by the oratory of Secretary of State Thomas Jefferson, who cautioned, "No invasions of the Constitution are fundamentally so dangerous as the tricks played on their own numbers, apportionment." Jefferson put forward a plan of his own, which Washington adopted despite its serious drawback of violating quota.

In Balinski and Young's five-state example, each house member would ideally represent 1,000 people, since the total population (26,000) divided by the House size (26) is 1,000. Hamilton's method has the effect of dividing each state's population by 1,000 and then rounding down for all states except those

---

*Balinski and Young, "The Quota Method of Apportionment," *American Mathematical Monthly* 82 (August–September 1975): 701–30.

with the largest fractions, which are rounded up as needed to fill out the house. Instead of using the divisor 1,000, Jefferson's method (also known as the method of greatest divisors) calls for using the largest divisor that will yield numbers for each state that when left alone or rounded down sum to the size of the house. In other words, the numbers never need to be rounded up. In the five-state example, 906.1 turns out to be the greatest divisor that gives such a result:

State	Population	Hamilton Divisor of 1,000 for 26 seats	Hamilton Allotment	Jefferson Divisor of 906.1 for 26 seats	Jefferson Allotment
A	9,061	9.061	9	10.000	10
B	7,179	7.179	7	7.923	7
C	5,259	5.259	5	5.804	5
D	3,319	3.319	4	3.663	3
E	1,182	1.182	1	1.304	1
Total	26,000		26		26

As the above table shows, Jefferson's and Hamilton's methods yield different results. Under Jefferson's, state A—the most populous state—gains a representative (and state D loses one). That Jefferson's method helps state A is no fluke; it can be shown mathematically that it favors large states. His lofty oratory never addressed this mathematical favoritism, although, being a shrewd man of science, he was no doubt fully aware of it. But it was a favoritism he approved of, since he was from the largest state, Virginia (population 630,558), as was Washington. Indeed, in the first apportionment of House members, in 1792, Jefferson's method (as opposed to Hamilton's) ensured Virginia an additional representative, at the expense of the tiniest state, Delaware (population 55,538).

Jefferson's method was followed more or less for half a century, from 1792 until 1841. (I say "more or less" because sometimes the House size was not fixed in advance but was adjusted, in the interests of political expediency, so that states would not lose representatives under a new apportionment.) Daniel Webster, recognizing that Jefferson's method underrepresented the New England states, Webster's home turf, persuaded Congress to adopt a new apportionment scheme. Like Jefferson's method, Webster's (also called the method of major fractions) is based on the selection of a greatest divisor. But the resulting numbers are not automatically rounded down but rounded according to the standard convention, down for fractions of less than .5 and up for fractions of .5 and above. For the five states, the greatest such divisor is 957.2, and state B does better than it did in either of the other methods:

State	Pop.	Webster Divisor of 957.2 for 26 seats	Webster Allotment	Hamilton Allotment	Jefferson Allotment
A	9,061	9.466	9	9	10
B	7,179	7.500	8	7	7
C	5,259	5.494	5	5	5
D	3,319	3.467	3	4	3
E	1,182	1.235	1	1	1
Total	26,000		26	26	26

Each step of the way, a few congressmen argued against upping the House's size, but their appeals, however persuasive, fell on deaf ears. Curiously, much more was made of the unwieldiness of a larger House than of the illegality; the remarks of Representative Samuel Cox of New York were typical: "A body is not great by being big. Corpulence is not health or rigor. A wheezy adiposity is not necessarily a condition of mental alertness. Layers of lard and monstrosities of fat are not conducive to manhood."

The failure to follow Hamilton's method was of no small consequence: in 1876, it robbed Samuel Tilden of the presidency. In the electoral college, each state receives a number of electors equal to the number of its representatives and senators. In that famous election, Tilden received 264,292 more popular votes than Rutherford B. Hayes but lost the election when Hayes received one more electoral vote than Tilden. Balinski and Young demonstrate that if Hamilton's method had been followed, as the law required, Tilden would have won, because one of the states he carried should have had an additional elector, at the expense of a state that Hayes carried.

The Alabama paradox was finally detected in 1881, when the chief clerk of the census office was investigating, on the basis of the 1880 census, various apportionments for House sizes ranging from 275 to 350. "While making these calculations," the clerk wrote to a member of the House, "I met with the so-called 'Alabama paradox' where Alabama was allotted 8 representatives out of a total of 299, receiving but 7 where the total became 300." For another twenty years, however, the Alabama paradox remained a defect more in theory than in practice.

Then, in 1901, when House seats were being reapportioned on the basis of the census of 1900, the Alabama paradox became a practical problem, evoking vitriolic debate. The majority of the House pushed through a bill that set the size of the House at 357 seats, of which Colorado received 2. Denouncing "the atrocity which [mathematicians] have elected to call a paradox," Representative John C. Bell of Colorado observed that in every other size House from 350 to 400, his state would receive not two but three representatives. In the 357-seat House, Maine also suffered from the Alabama paradox, and one of its representatives said, "It does seem as though mathematics and science have combined to make a shuttlecock and battledore of the State of Maine. . . . God help Maine when mathematics reach for her!"

Over the next few decades, eminent mathematicians paraded before the House and offered sophisticated numerical formulas, incomprehensible to most of the politicos, for avoiding the Alabama paradox. One of these formulas was adopted in 1941, when Franklin Roosevelt signed "An Act to Provide for Apportioning Representatives in Congress among the several States by the equal proportions method."

The method of equal proportions had been proposed twenty years earlier by Edward V. Huntington, a Harvard mathematician. Huntington argued that, given the reality of so many different states with different populations, when the representation granted any two states is compared, one of the states will inevitably be shortchanged and that the amount of shortchangedness can be measured. If the transfer of one representative from the better-off state to the worse-off state reduces the relative amount of shortchangedness, the transfer should be made. For example, if in a comparison of Virginia and Massachusetts, Virginia is found to be worse off by 3 units of shortchangedness and the transfer of one representative from Massachusetts to Virginia turns the tables in such a way that Massachusetts is now worse off by 2 units, the transfer should be made, because the relative amount of shortchangedness—2 units as opposed to 3—is reduced. If, instead, the transfer turns the tables to such an extent that Massachusetts is worse off by 4 units, the transfer should not be made, because the status quo is more equitable. The idea is to apportion representatives in such a manner that the relative amount of shortchangedness is minimized. This will occur when no pairwise comparison of states dictates the transfer of a representative.

The idea of trying to minimize the relative amount of short-changedness is appealing, but how is shortchangedness to be measured? In the method of equal proportions, one computes shortchangedness by first taking the numerical difference between the average size of a state's congressional district and the average size of another state's congressional district and then expressing that difference as a fraction of the smaller district size. In the five-state example, the method of equal proportions yields yet another assignment of representatives, in which state C benefits:

State	Population	Average District Size	Equal Proportions Allotment for 26 seats
A	9,061	1,006.78	9
B	7,179	1,025.57	7
C	5,259	876.50	6
D	3,319	1106.33	3
E	1,182	1182.00	1
Total	26,000		26

According to the measure of shortchangedness described above, state D is shortchanged by $(1106.23–876.50)/876.50$, or .2621. The transfer of one representative from D to C would change the average district size in state C to 1,051.80 and in state D to 829.75. This allotment is less equitable because the relative amount of shortchangedness is increased, state C being shortchanged by $(1051.80–829.75)/829.75$ or .2676. If you play around with the numbers in the above table, you'll find that, given this measure of shortchangedness, no other assignment of representatives is more equitable.

This measure of shortchangedness, however, has no a priori claim to being fair. You might, instead, simply compute the difference between the two sizes and not bother to express it as a fraction. Or you might compute for each state what fraction of a representative each resident corresponds to and try to minimize the difference in these fractions from state to state. There are other possibilities, too, all with equal claims to being fair.

The problem in defining shortchangedness can be understood by analogy. Suppose I tell you that Bob's annual income exceeds Jake's by $10,000. By this measure—the absolute difference in income—Jake is $10,000 worse off, but this doesn't tell you everything you might want to know in evaluating their relative standard of living. Jake might make only $10,000 a year, in which case Bob is earning 100 percent more. But Jake could also be raking in $1,000,000 a year, in which case Bob is earning only 1 percent more. If the difference in income were reported not in terms of absolute dollars but in terms of percent, other information you might need in order to judge their standard of living would be suppressed. For example, suppose you know that Bob earns 100 percent more money than Jake. That doesn't tell you whether Bob could live just as well as Jake and, in addition, buy a $100,000 house in cash. If Bob earns $200,000 (to Jake's $100,000), he'd have the extra cash to spare. But if he makes only $10,000 (to Jake's $5,000), he'd have to settle, say, for a home computer instead of a house. The moral is that no measure of income disparity—be it absolute dollars, percentage difference, or something else—has an a priori claim to being the best measure. The same is true of measuring the relative extent to which states are shortchanged in their delegations to the House.

Unbeknownst to Roosevelt and the Congress, the method of equal proportions also violates quota, as Balinski and Young noted in the *American Mathematical Monthly*. Moreover, it tends to favor smaller states. (You can discern either of these drawbacks from the five-state example.) Perhaps you're beginning to get the idea that every apportionment system is plagued by paradox, aside from the obvious inequalities resulting from the inability to divide a congressman. In their 1982 book, *Fair Representation: Meeting the Ideal of One Man, One Vote*, Balinski and Young offer a mathematical proof that no apportionment method exists that always satisfies quota and always avoids the Alabama paradox.

In the best tradition of social-choice theory (the branch of applied mathematics that addresses how the preferences of individuals should be combined to get a social choice), Balinski and Young don't stop with the mere identification of paradoxes but go on to investigate how frequently they crop up. After all, the real world demands a solution—that representatives be apportioned by one method or another—and a method that is almost always free of paradoxes is clearly preferable to one riddled with them. Balinski and Young were able to show, on the basis of randomly generated population data, that Webster's method does not discriminate in favor of big or small states and that it is less likely to violate quota than are other apportionment methods immune to the Alabama paradox.

Will Balinski and Young's compelling analysis inspire a movement in Congress for a return to Webster's method? If the method were used today (instead of the method of equal proportions), the only difference would be that New Mexico would forfeit a seat to Indiana. In the House subcommittee on census and population, the Indiana delegation, armed with Balinski and Young's analysis, proposed a bill reinstating Webster's method. But the bill aroused little interest (except the ire of the New Mexico delegation) and died in subcommittee. Alas, the social-choice theorist's lot is a lonely one.

·

SUGGESTIONS FOR FURTHER READING

Numbers

Paul Hoffman. "The Man Who Loves Only Numbers." *The Atlantic,* November 1987, pp. 60–74. A profile of Paul Erdös, the world's most prolific mathematician and a consummate number theorist.

G. H. Hardy. *A Mathematician's Apology.* Cambridge University Press, 1940. A classic, stylish defense of pure mathematics by one of its leading practitioners.

Martin Gardner. *The Incredible Dr. Matrix.* Charles Scribner's Sons, 1976. Gardner's protagonist, Dr. Irving Joshua Matrix, is a recreational mathematician who spots numerical coincidences everywhere.

Shapes

Stefan Hildebrandt and Anthony Tromba. *Mathematics and Optional Form.* Scientific American Books, 1985. A beautifully illustrated book that is must reading for devotees of mathematical shapes.

Machines

Peter W. Frey. *Chess Skill in Man and Machine.* Springer Verlag, 1983. A series of essays probing how to program a computer to play strong chess.

Paul Hoffman. "A Chess Player Realizes the Game Controls His Life." *Smithsonian,* July 1987, pp. 129–140. An account of the emotional side of chess.

One Man, One Vote

Steven J. Brams. *Paradoxes in Politics.* The Free Press, 1976.
Steven J. Brams and Peter C. Fishburn. *Approval Voting.* Birkhauser, 1983. Clear presentations of game theory paradoxes in voting.

INDEX

Made in the USA
Lexington, KY
16 November 2011